PRAISE FOR *GARBOLOGY*

"This is a horrifying, well-documented, and fascinating study of how profligate waste became a normal part of American consumer behavior and what it's going to take for our society to shift from a disposable economy to a reusable one . . . This should be a "One Book" reading selection in every American community." — LIBRARY JOURNAL

"An eye-opening account reminding us of something we try to forget: We are a wasteful society with a trash problem that is polluting our oceans and packing our landfills." — THE BOSTON GLOBE

"[Humes] exhibits passion for his cause." — THE WALL STREET JOURNAL

"*Garbology* is [Humes's] attempt to make sense of our historically unprecedented readiness to throw things away . . . Food for thought, and more." — THE ECONOMIST

"Humes's take on the science and culture of 'garbology' is both academic and deeply personal, making this a fascinating read." — PUBLISHERS WEEKLY

"Zestful in his curiosity and irrepressible in his vivid chronicling . . . Humes finds hope in the innovative work of dedicated garbologists, trash trackers, and activists who are intent on exposing the hazards and travesties of excessive trash and pointing the way to the 'low-waste path.'" — BOOKLIST (starred review)

"Humes's argument isn't a castigation of litterbugs. It's a persuasive and sometimes astonishing indictment of an economy that's become inextricably linked to the increasing consumption of cheap, disposable stuff—ultimately to our own economic, political, and yes, environmental peril." — BOOKFORUM

"Humes offers plenty of surprising, even shocking, statistics . . . an important addition to the environmentalist bookshelf." —KIRKUS REVIEWS

"Unlike most dirty books, this one is novel and fresh on every page. You'll be amazed." —BILL MCKIBBEN, author of *Eaarth*

"Edward Humes takes us on a real romp through the waste stream. *Garbology* is an illuminating, entertaining read that ultimately provides hope and tips for a less wasteful future. This book will make you want to burn, or at least recycle, your trash can!"
—JONATHAN BLOOM, author of *American Wasteland*

"In this well-written and fast-paced book, Ed Humes delves into the underbelly of a consumer society—its trash. What he finds is so startling and infuriating, you will never think about 'waste' in the same way again."
—SAMUEL FROMARTZ, author of *Organic, Inc.* and editor-in-chief of the Food & Environment Reporting Network

GARBOLOGY

OUR DIRTY LOVE AFFAIR WITH TRASH

EDWARD HUMES

AVERY
a member of Penguin Group (USA) Inc.
New York

Published by the Penguin Group
Penguin Group (USA) Inc., 375 Hudson Street,
New York, New York 10014, USA

USA · Canada · UK · Ireland · Australia
New Zealand · India · South Africa · China

Penguin Books Ltd, Registered Offices: 80 Strand, London WC2R ORL, England
For more information about the Penguin Group, visit penguin.com

First trade paperback edition 2013
Copyright © 2012 by Edward Humes

Published simultaneously in Canada

Most Avery books are available at special quantity discounts for bulk purchase
for sales promotions, premiums, fund-raising, and educational needs. Special books
or book excerpts also can be created to fit specific needs. For details, write
Penguin Group (USA) Inc. Special Markets, 375 Hudson Street, New York, NY 10014.

The Library of Congress has catalogued the hardcover edition as follows:

Humes, Edward.
Garbology : our dirty love affair with trash / Edward Humes.
p.   cm.
Includes bibliographical references.
ISBN 978-1-58333-434-8 (hardback)
1. Refuse and refuse disposal—United States.   2. Environmental engineering—
United States.   3. Salvage (Waste, etc.)—China.   I. Title.
TD788.H86   2012   2012001701
628.4'40973—dc23

ISBN 978-1-58333-523-9 (paperback edition)

Printed in the United States of America
5   7   9   10   8   6   4

BOOK DESIGN BY TANYA MAIBORODA

In memory of my grandmother Maggie,
who survived famine, weathered the Great Depression,
drank her Irish whiskey neat, taught me to play poker at age seven,
and instructed me that, while wasting is not exactly a sin,
it is rather stupid

CONTENTS

PART 3. THE WAY BACK

INTRODUCTION: 102 TONS (OR: BECOMING CHINA'S TRASH COMPACTOR)

On May 24, 2010, rescue workers donned impermeable hazardous materials suits, then burrowed into the creaking, dangerous confines of a ruined South Side Chicago home, searching for the elderly couple trapped inside.

More than an hour later, as curious neighbors gathered and a television news crew arrived to film the emergency rescue operations, Jesse Gaston, a seventy-six-year-old chemist, and his wife, Thelma, a retired schoolteacher, walked unsteadily into the hazy afternoon light, dehydrated and hungry but still among the living.

The Gastons had been trapped by trash—their own trash.

The debris had accumulated for years until every surface of the house was covered by layers of old newspapers, empty plastic jars, pieces of broken furniture, worn-out coolers, splintered garden rakes, thousands of soda bottles, cans of every size, clothing old and new, broken lamps, dusty catalogs, mountains of junk mail and garbage bags filled with the detritus of daily life. All of this, and much more, had been kept for reasons no one could coherently explain, not even the Gastons, until the junk and trash reached the level of the highest kitchen cupboards, the ones that held the good china. A broken refrigerator lay in the kitchen, half buried and resting on its side, as if buoyed up by the sea of bottles, cans, cartons and sacks engulfing it. No room in the house could be called usable or even safely navigable; the stairs were blocked, the furniture buried, the garage packed floor to ceiling. The disordered accumulation looked as if it had been swept in by a tidal wave.

The Gastons simply grew unable to part with their trash. This hoarding compulsion gripped them gradually, a slow evolution, a piece at a time, then a bag here and there, then whole boxes of trash until, finally, the Gaston home became a one-way depository, the garbage version of the Eagles' famous "Hotel California": stuff checked in, but it could never leave. They hoarded until goods and trash consumed their home and almost their lives. Neighbors, alarmed by the fact that the couple hadn't been seen in three weeks—not to mention the increasingly persistent stench emanating from the home—had called police and firefighters. The rescuers eventually determined that Thelma had become trapped by falling debris somewhere in the bowels of the house, and Jesse, trying to reach her, had been pinned by piles of trash that toppled around him, too.

Although most of us tend to view this sort of extreme hoarding as an aberration, it's a surprisingly common occurrence. Variations of the Gaston household are found around the country more or less on a daily basis, although most often after the hoarder's demise, and seldom with the fanfare of news coverage. Somewhere between 3 and 6 million Americans are thought to be compulsive junk hoarders with living spaces that, to varying degrees, resemble the Gastons'. The phenomenon offers enough freak-show fascination to have spawned a cable television series: the A&E Network's *Hoarders*, which entertains viewers by taking them inside different hoarders' homes every week. The show's website offers a handy interactive quiz to help viewers determine if they, too, are addicted to hoarding or merely "just messy."

This phenomenon has not yet achieved true immortality as a distinct mental illness—the bible of psychiatry, the *Diagnostic and Statistical Manual of Mental Disorders*, categorizes extreme hoarding as merely one of many forms of the catch-all obsessive-compulsive disorder—although some experts are lobbying to have it classified as its own, unique ailment: disposophobia. The proposed malady is alternatively known as Collyer Brothers syndrome, named for one of the earliest and most dramatic manifestations of media-immortalized trash hoarding. Homer and Langley Collyer, rich and reclusive, rebelled early in the twentieth century against the still-evolving practice of mandatory municipal garbage collection in New York City. They turned their multistory brownstone into a crammed and putrefying museum of trash, featuring endless piles and rows of newspapers, bottles, boxes, broken gadgetry (Langley Collyer fancied himself an inventor) and a partially buried Model T Ford hidden on the second floor beneath layers of debris. The brothers were found dead in their home in 1947—Langley had been crushed by a collapsing tunnel of trash, and his invalid older brother, Homer, helpless without Langley's care, died later of thirst and starvation. Authorities eventually removed about 130 tons of trash from the brothers' home.

As morbidly compelling as such extreme hoarding may be (healthy ratings for cable television's looky-loo show earned it a multiple-season renewal deal, along with spawning a rival program, TLC's *Hoarding: Buried Alive*), the most revealing aspect of disposophobia is society's tone-deaf response to the phenomenon. The focus of therapists, "organization coaches," family, friends and TV show hosts is always on persuading disposophobics to do as "normal" people do: take the trash to the curb so it can be hauled away. A little counseling here, a little psychoactive drug therapy there, throw in a cleanup crew, a dump truck and some liberal doses of Mr. Clean and, poof, problem solved. But little if any thought is given to the refuse itself, or to the rather scarier question of how any person, hoarder or not, can possibly generate so much trash so quickly.

Of course, there's a reason for this blind spot: namely, the amount of junk, trash and waste that hoarders generate is perfectly, horrifyingly normal. It's just that most of us hoard it in landfills instead of living rooms, so we never see the truly epic quantities of stuff that we all discard. But make no mistake: The two or three years it took the Gastons to fill their house with five to six tons of trash is typical for an American couple. The Collyer brothers were outliers in their own time, but they would fit in the normal range circa 2011 quite nicely: Their lifetime trash production seventy years ago matches almost to the pound the prodigious modern American equivalent. The rest of us are just better at hiding it— mostly from ourselves.

This turns out to be something various government, industry

and university surveys attempt to track quite carefully: Americans make more trash than anyone else on the planet, throwing away about 7.1 pounds per person per day, 365 days a year.[1]

Across a lifetime that rate means, on average, we are each on track to generate 102 tons of trash.

Each of our bodies may occupy only one cemetery plot when we're done with this world, but a single person's 102-ton trash legacy will require the equivalent of 1,100 graves. Much of that refuse will outlast any grave marker, pharaoh's pyramid or modern skyscraper: One of the few relics of our civilization guaranteed to be recognizable twenty thousand years from now is the potato chip bag. (And no, those new biodegradable plastic bottles and bags intended to save the day so far haven't saved much of anything. Turns out manufacturers failed to check whether their lab-tested degradability is compatible with the real-world network of local composters and recyclers across the nation. Mostly, they're not.[2])

And so the trash trail only grows: The Environmental Protection Agency estimates that, between 1980 and 2000, the average American's daily trash load increased by a third. The difference between now and 1960 is even greater, at least double the per capita trash output. Americans have "won" the world trash derby without really trying, making at least 50 percent more garbage per person than other Western economies with similar standards of living (Germany, Austria and Denmark, among others), and between two and three times the trash output of the Japanese. America's production of waste exceeds past projections of previous generations who tried to estimate how wasteful their twenty-first-century counterparts would be. The futurist marketers behind the 1964 World's Fair in New York felt they were being fairly conservative when they built scale models of the gleaming future cityscapes we were supposed to be living in by now (hover cars and moving sidewalks, anyone?) in which problems of energy and waste had been solved by technology rather than exacerbated by it. Garbage was so old school; we were supposed to have scienced away that ancient problem ages ago.

What no one considered back then (and few acknowledge now) is waste's oddest, most powerful quality: We're addicted to it.

It turns out our contemporary economy, not to mention the current incarnation of the American Dream, is inextricably linked to an endless, accelerating accumulation of trash. The purchases that drive the markets, the products that prove the dream, all come packaged in instant trash (the boxes, wrappers, bags, ties, bottles, caps and plastic bubbles that contain products). And what's inside that packaging is destined to break, become obsolete, get used up or become unfashionable in a few years, months or even days—in other words, rapidly becoming trash, too. When the tide of garbage bound for the landfill grows from year to year, America's leaders rejoice because, despite the economic and environmental cost of waste, it signals the welcome reality that more people and businesses are buying more stuff. This is why countries with booming economies—China being the prime example—are frantically digging new landfills to ring their growing cities.

Garbage has become one of the most accurate measures of prosperity in twenty-first-century America and the world.

The opposite holds true as well. When the lines of garbage trucks headed to America's landfills grow shorter, as they did in 2008 and the years that followed, it makes for a surer sign that our disposable economy is headed for recession than a plunging Dow Jones Industrial Average or a falling dollar. No stockbroker could out-predict the landfill dozer and compactor operators, who saw the bubble bursting ahead of everyone. Presidents used to fret that Americans did not save enough. Now they worry when we do not shop enough, the modern cure for recession and economic crisis, epitomized by President George W. Bush's call to Americans in the wake of the 9/11 attack to go out and spend more money for the good of the country. This prevailing viewpoint that favors spending rather than saving our way to prosperity, whatever its merits, creates a powerful societal and economic undertow that fuels America's garbage addiction.

It's an ailment that did not exist in anything like its current form for 99.9 percent of human history. Today's hoarders perform a kind of public service, letting us see what our true legacy looks like. Otherwise those 102 tons remain virtually invisible, too big to see. We chuck pieces of it in the can every day, push it out to the curb every week and then forget about it as if it's gone. But that clever vanishing trick hides the fact that nothing people do has more impact than their waste. It's connected to everything: energy, food, pollution, water, health, politics, climate, economies. Trash is nothing less than the ultimate lens on our lives, our priorities, our failings, our secrets and our hubris.

One out of every six big trucks in the U.S. is a garbage truck. Their yearly loads would fill a line of trucks stretching halfway to the moon. The creation of products and packaging that end up in those trucks contributes 44 percent of the greenhouse gas emissions that drive global warming, more than any other carbon-spewing category.[3] Garbage costs are staggering: New York City alone spent $2.2 billion on sanitation in 2011. More than $300 million of that was just for transporting its citizens' trash by train and truck—12,000 tons a day—to out-of-state landfills, some as far as three hundred miles away. How much is 12,000 tons a day? That's like throwing away sixty-two Boeing 747 jumbo jets daily, or driving 8,730 new Honda Civics into a landfill each morning. Imagine an armada of the U.S. Army's mighty M-1 Abrams main battle tanks lined up bumper to bumper for more than a mile. That's 12,000 tons—one city's trash, one very costly day.

Now multiply all that about thirty-six times to gauge the nation's daily garbage spend and flow. In a year, Americans throw out a collective 389.5 million tons of rubbish—what the feds call "municipal solid waste,"[4] the stuff we personally throw away. This annual load of trash is roughly equivalent to the collective weight of the entire U.S. adult population—eighteen times over.[5]

This staggering number is not easy to find, because like any addict, America is living in an official state of garbage denial. The

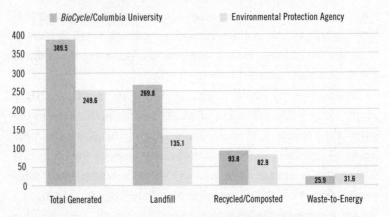

How Much Trash We Make, by Millions of Tons: The "Real" vs. the "Official" Numbers, 2008

Millions of tons of municipal solid waste, by destination, comparing the BioCycle/Columbia University physical count of trash tonnage vs. the Environmental Protection Agency's theoretical "materials flow analysis." The EPA is reassessing its methods. Data for 2008 is displayed, the last year covered by both measures.

Sources: "Municipal Solid Waste in the U.S.," EPA, 2009; "The State of Garbage in America," BioCycle, October 2010[6]

Where Our Trash Goes: The "Real" vs. the "Official" Numbers

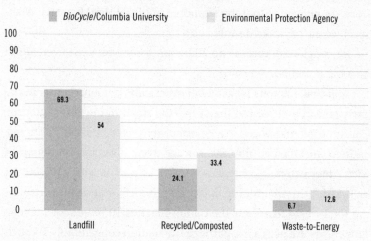

Percentage breakdown of municipal solid waste, comparing the BioCycle/Columbia University physical count of trash tonnage vs. the Environmental Protection Agency's theoretical "materials flow analysis." Percentages displayed are for the year 2008, the most recent available for both measures. The EPA is reassessing its methods.

Sources: "Municipal Solid Waste in the U.S.," EPA, 2009; "The State of Garbage in America," BioCycle, October 2010[7]

statistical bible of municipal waste put out annually by the Environmental Protection Agency—accepted for decades as the garbage gold standard by policy makers and media alike—scandalously underestimates America's trash by relying on byzantine simulations and equations rather than actual counts of trash going to landfills. More than 140 million tons of garbage come up unaccounted for in the process. It turns out that obscure but far more accurate scientific surveys made jointly by Columbia University and the journal *BioCycle* reveal that we're sending twice as much waste to landfills as the EPA's calculations let on, and recycling proportionately far less than the rosy official stats suggest. The EPA reports a third of our trash gets recycled or composted, but the real-world figures indicate that this diversion rate is less than a fourth of our total trash—a milestone that the supposed gold standard incorrectly asserts we surpassed a decade ago.[8]

It's tough to overcome an addiction when you can't even admit how big a problem you've got.

And that 102 tons is just what Americans personally toss in the garbage can and haul to the curb—the trash in our direct control. Counting all the waste transported, extracted, burned, pumped, emitted and flushed into the sewage system by and on behalf of each American man, woman and child, as well as what's tossed out by U.S. industry in order to make the products Americans consume, the total waste figure for the nation reaches 10 billion tons a year. This raises the per capita garbage calculation considerably. By such an all-waste accounting, every person in America stands atop more than 35 tons of waste a year—or a staggering average lifetime legacy of 2,700 tons. No wonder America, with 5 percent of the world's population, accounts for nearly 25 percent of the world's waste.

Then there's the wallet issue. Trash is such a big part of daily life that American communities spend more on waste management than on fire protection, parks and recreation, libraries or schoolbooks. If it were a product, trash would surpass everything else we manufacture. And guess what? It *has* become a product—America's leading export.

That's the secret behind the story of Zhang Yin, another sort of hoarder, one who is admired rather than pitied. In 2006, she became at age forty-nine China's first woman billionaire. In 2011, she was both China's top female manufacturer and America's biggest exporter *to* China (of either gender). Her export: America's garbage. In both East and West, she is the queen of trash.

Zhang is also the personification of the American Dream in

the twenty-first century, a Horatio Alger for a disposable economy. Fleeing the Tiananmen Square massacre and democracy movement crackdown of 1989, she left China for the Los Angeles suburb of Pomona, where she started running a scrap-paper company out of her apartment. The entire workforce at first consisted of Zhang and her new husband, a Taiwanese immigrant trained as a dental surgeon. They would drive around the Los Angeles Basin in an old Dodge van, visiting landfills and their sorting and recycling stations, begging for scrap paper. Learning English as she built the business, Zhang cut a series of deals to secure a steady source of the waste paper at a bargain price. There was no shortage of material. Then, as now, paper waste was one of the main components of trash dumped at landfills. American businesses considered much of the material worthless.

China, on the other hand, had a chronic paper and pulp shortage, having deforested huge swaths of the country during the drive to industrialize in the late fifties and early sixties—"the Great Leap Forward," as it was called. In the nineties, as manufacturing ramped up and China joined the global economy in earnest, there was enormous demand for cardboard to package and box the goods that China had begun to produce. The scrap paper Zhang amassed was just what the Chinese factories needed—they'd recycle all she could send them. Because cargo ships were coming to America from China full and returning mostly empty, Zhang was able to negotiate bargain-basement shipping costs to her native land.

Soon she had deals with recyclers and brokers all over Los Angeles, New York and Chicago to fill the voracious demand. "Chinese manufacturers were desperate for scrap paper," she recalled years later. "I'm an entrepreneur . . . All I did was help fulfill a need."

That's probably a bit too modest. The daughter of a Red Army officer imprisoned during China's Cultural Revolution, she managed to see an opportunity that American entrepreneurs had missed. She filled China's paper needs so thoroughly that, beginning in the year 2000 and every year since, her company, America Chung Nam, has been the top U.S. exporter to China in number of cargo containers shipped—and the largest scrap-paper company in the world, an empire of trash built from scratch. She used the earnings—and America's scrap—to launch what is now China's largest cardboard manufacturer, Nine Dragons Paper; by 2010, she was worth $4.4 billion.

Zhang is a big part of a simple but rarely acknowledged fact about America's place in the twenty-first-century global economy: Trash has become one of the most prized products made in the USA. Not computers. Not cars. Not planes or missiles or any other manufactured product. It's our mountains of waste paper and soiled cardboard and crushed beer cans and junked electronics that the rest of the world covets.

In 2010, China's number one export to the U.S. was computer equipment—about $50 billion worth.[9] America's two highest volume exports to China were paper waste and scrap metal, a little more than $8 billion worth of bundled old newspaper, crushed cardboard, rusty steel and mashed beverage cans sold at rock-bottom prices. Zhang's America Chung Nam exported more than three hundred thousand cargo containers of scrap paper to China in 2010. Overall, the fastest-growing category of goods exported to China is "Scrap and Trash," increasing 916 percent between 2000 and 2008.[10] Chinese manufacturers promptly develop new and aggressively priced consumer products made from this waste, which they then sell back to American consumers at great profit, so we can trash it all again in a year or two and send it back once more for pennies on the dollar. Waste, it seems, is becoming one of our greatest contributions to the global economy.

Somehow, without ballot or poll or any explicit decision by presidents or legislators or voters to do so, America, a country that once built things for the rest of the world, has transformed itself into China's trash compactor.

This sobering economic reality is mirrored by a telling observation from, of all sources, America's astronaut corps: There are only two man-made structures large enough to be identifiable without magnification from earth orbit. First, there's the mighty Great Wall of China in the East, symbol of a past power risen again. And in the West, there's a newer thing, the grimly named Fresh Kills, recognizable above all other things American.

Fresh Kills is the world's largest town dump, the recently shuttered repository for a half century's accumulation of New York City garbage.

ANY ATTEMPT to understand the 102-ton legacy—and what can (or should) be done about it—has to begin with answers to three very basic (yet rarely posed) questions. As it happens, these are the same three questions extreme hoarders such as the Gastons must confront if they wish to change their trash-laden circumstances:

First there's the most obvious of inquiries: What is the nature and cost of that 102-ton monument of waste?

Next comes the soul-searching question: How is it possible for people to create so much waste without intending to do so, or even realizing they are doing it?

Finally, there's the "what next?" question: Is there a way back from the 102-ton legacy, and what would that do for us . . . or to us?

Problem, investigation, solution: It's the classic three-act construction that the human brain has been hardwired to prefer—and as good an organizing principle as any for a book about trash. Three sections, three broad questions, each equally important, but it's the third piece of the story, the quest for a way back, that is key. That's the question that allows the 102-ton story to become a voyage of discovery, offering the possibility that all those tons of garbage might be a choice rather than an inevitability—and an opportunity as well as a bane. That's the question that offers the possibility of a happy ending to the story of trash.

Oddly enough, it's the hoarders, once again, who can help show us the way back. The Gastons understood far better than their neighbors that our prevailing definition of waste is all wrong. They saw that putting something in the trash is not really a matter of disposing of waste, of something with no value. Trash to them is the physical manifestation of *wastefulness*. The hoarders' response to this essential insight—that trash is really treasure squandered—is twisted and unhealthful, but their instinct to place value on garbage is sound and sane. Of course, the more constructive response would be not to hoard, but to find ways to avoid the wasteful accumulation in the first place. That's the great challenge, the holy grail that has so far eluded mankind, dating all the way back to the first town dump and anti-littering law in ancient Greece. The upside of this picture: There is a small but growing number of businesspeople, environmentalists, communities and families who see in our trash the biggest untapped opportunity of the century.

These trash optimists range from the city of Portland, which may be the least wasteful city in America, to TerraCycle, the business champion of "upcycling" (the reuse/repurpose opposite of recycling), to the trash artists of San Francisco and the trash czar at

Harvard University who each year turns the stuff students abandon in the dorms into one of the biggest and most successful yard sales in America. And there's the Johnson family, who proved they could live an outwardly normal year and yet produce only a mason jar full of trash.

Bea Johnson, a Marin County, California, artist who set her family of four on this quest, wonders what would have happened if the massive infrastructure America has constructed to deal with trash had been predicated all along on avoiding waste and recapturing its value, rather than transporting, burying and occasionally recycling its epic quantities. Would America still be evolving into China's trash compactor? Would there even be a 102-ton legacy? "What would life look like then?" she muses. "What would it mean for the economy, for the entire world?"

Johnson (you'll read more about her trash epiphany later) is the opposite of a hoarder—she's all about avoiding the accumulation of things, particularly disposable things, and living the uncluttered life. Or as she calls it, the unwasteful life. She says people, even friends, question her sanity, but the Johnson family has discovered that generating less waste translates into more money, less debt, more leisure time, less stress. When they give gifts, they don't give things—they give experiences. No wrapping paper required. She says they've never been happier.

"When you stop wasting, everything changes," she says. "There is a way back. And if it can work for a family, it can work for a country. It could be the answer we've all been waiting for."

AN AMERICAN ANNUAL WASTE SAMPLER

- 5.7 million tons of carpet sent to landfills—all of it could be recycled, but mostly it's not
- 19 billion pounds of polystyrene peanuts (Styrofoam) dumped—never degrades, impossible to recycle
- 35 billion plastic bottles
- 40 billion plastic knives, forks and spoons
- 4.5 million tons of office paper
- Enough aluminum to rebuild the entire commercial air fleet four times over
- Enough steel to level and restore Manhattan
- Enough wood to heat 50 million homes for twenty years
- Enough plastic film to shrink-wrap Texas
- Plastic waste is so plentiful and so carelessly treated that 92 percent of Americans have potentially harmful plastic chemicals in their urine
- 10 percent of the world oil supply is used to make and transport disposable plastics
- Growing, shipping and selling food destined to be thrown away uses more energy than is currently produced by offshore oil drilling
- No less than 28 billion pounds of food thrown away, about 25 percent of the American food supply, perhaps more by some estimates

PART

1

THE BIGGEST THING WE MAKE

Our willingness to part with something before it is completely worn out is a phenomenon noticeable in no other society in history . . . It must be further nurtured even though it runs contrary to one of the oldest inbred laws of humanity, the law of thrift.

—J. GORDON LIPPINCOTT, 1947

A society in which consumption has to be artificially stimulated in order to keep production going is a society founded on trash and waste, and such a society is a house built upon sand.

—DOROTHY L. SAYERS, 1947

Who steals my purse steals trash.

—IAGO, IN SHAKESPEARE'S *Othello*

1

AIN'T NO MOUNTAIN
HIGH ENOUGH

MIKE SPEISER'S SCULPTING TECHNIQUE IS A STUDY in geometric perfection and economy of motion. Every cut, every shave, every subtle drag of his blade has a purpose, each forming a small piece of a much larger work, sprawling and unique.

His peers call him Big Mike, for he is a mountain of a man, shaved head set like an amiable boulder atop broad shoulders and a mighty belly, six-two and more than three hundred pounds. He seems designed by central casting exactly for the purpose of wielding his main artistic tool—the towering, thundering 60-ton BOMAG Compactor. With its roaring, clanking assistance, Big Mike has helped build something unprecedented: the Puente Hills landfill, largest active municipal dump in the country.

Puente Hills is so sprawling that it has evolved its own ecosystem and nature preserve, spawned multiple community organizations formed to kill it, and holds enough strata of methane-spewing decomposing garbage to power a hundred thousand homes (which is exactly what is done with the eye-watering "landfill gas" bubbling up from beneath). Puente Hills has been the final resting place for the lion's share of Los Angeles County's ample daily flow of garbage for more than three decades—130 million tons of it and counting.

One hundred thirty million tons: Such a number is hard to grasp. Here's one way to picture it: If Puente Hills were an elephant burial ground, its tonnage would represent about 15 million deceased pachyderms—equivalent to every living elephant on earth, times twenty. If it were an automobile burial ground, it could hold every car produced in America for the past fifteen years.

It is, quite literally, a mountain of garbage.

Big Mike's German-made BOMAG is the primary tool for taming this garbage nexus. The BOMAG (derived from the German-language mouthful of a company name, Bopparder Maschinenbau-Gesellschaft) is a fourteen-foot-tall, thirty-foot-long, swivel-hipped bulldozer that can turn on a dime yet push its terrain-clearing blade with 100,000 pounds of force. Its six-foot wheels are spiked with dinosaur-sized steel teeth that can crush, mold and squeeze up to 13,000 tons of garbage into a fifteen-foot-deep rectangle the length and width of a football field.

Big Mike sculpts such a mound not in a month or a week, but in one glorious day, every day, as he and his colleagues dump, push, carve and build a pinnacle of trash where once there were canyons. He is king of a mountain built of old tricycles and bent board games, yellowed newspapers and bulging plastic bags, sewage sludge and construction debris—all the detritus, discards and once valuable tokens of modern life and wealth, reduced to an amorphous, dense amalgam known as "fill."

The football-field-sized plot at the center of activity atop Puente Hills is called a "cell," not in the prison-block sense, but more akin to the tiny biological unit, many thousands of which are needed to create a single, whole organism. As with living creatures, this cell, titanic as it is, represents a small building block for the modern landfill—the part that grows and reproduces each day. A dozen BOMAGs, bulldozers and graders swarm over this fresh fill every day, backing and turning and mashing and shaping, their warning gongs clanging and engines roaring in a controlled chaos, mammoth bees crawling atop the hive. Their curved steel blades raise up and blot the sun, then drop into the sea of trash and push it forward, waves of debris flowing off either side as if the dozers' blades were the prows of a schooner fleet, complete with the flap and quarrel of seagulls overhead, their cranky squawks drowned out by the diesel din. A sickly-sweet smell of decay kicks up when the cell is churned this way, and the thrum and grind of the big engines can be felt in the ground near the cell. The noise induces sympathetic vibrations in the chest of anyone nearby, creating the uncomfortable sensation of being near a marching band with too many bass drummers.

Building a garbage mountain is difficult, edgy, dangerous work. Within the new cell, the trash flow can pile up twenty to thirty feet or more during the day before it's crushed and compacted and covered with clean dirt (that's what makes it a *sanitary* landfill—the ick gets buried every day). The drivers negotiating and moving that cell-in-the-making must constantly be wary of the drop-off from their garbage pile—and the uncontrolled, possibly tumbling sled ride that tipping over the edge could bring about. Eight landfill workers nationwide died on the job in 2010.

To build a proper trash mountain, one that is a feat of engineering rather than a random aggregation of garbage, each cell must be level at the top so it can be covered and sealed with up to a foot of soil, the last task of any day at the landfill. The machine operators rely on laser-guided markers to keep their mound level, except for Big Mike, who seems to be able to do it by dead reckoning alone. His coworkers say he can visualize the final form of a field of compacted trash the way an artist can see the carving within the block of wood or the figure hiding inside the marble. A member of Puente Hills's team of waste engineers, guys with hard hats and clipboards who plot out each day of garbage burial with the same care and planning once lavished on an Egyptian mummy's tomb, glances one morning at a section of new fill and says, "Oh, look at that perfect edge—that's Big Mike's work. That's his style." The other engineers nod.

Later, Mike grins sheepishly when he hears about the compliment. He's forty-eight and has been doing this for twenty years. The little fang earrings he sports jiggle a little with his chuckle. "I do love my work," he says. "Where else can you accomplish something every day, see the progress right in front of you, and know you're doing something useful and good? And on top of that, it's fun. Where else are you going to drive a hundred and twenty thousand pounds of machine around all day and get paid for doing it?"

His life's work is the mother of all landfills, its innovations and pioneering techniques copied and studied. But in truth, calling it a landfill these days is a bit misleading, as it stopped physically "fill-ing" a depression in the land (the original definition of landfill) long ago. For quite some time, the garbage mountain of Puente Hills has been rising above its surrounding terrain, resembling nothing more than a huge and eerily modern version of an ancient tell—those giant mounds in the Middle Eastern deserts that mark where once-mighty cities rose and fell, and that now lie buried and broken beneath the sands.

Archaeologists love tells—and, particularly, the middens they usually conceal, those ancient trash dumps that, five thousand years later, provide a treasure trove of information about life and events in the distant past. Archaeologists long ago figured out that the real nature of human life isn't that we are what we eat. They know we are best understood by what we throw away. Thousands of years from now, the Puente Hills landfill, buffered, insulated, wrapped in layers of clay and polyethylene, and more secure against earthquakes, winds and floods than any other structure in Califor-nia, may serve a similar archaeological purpose, a tell for future researchers hoping to puzzle out our lifestyle, choices and beliefs. Certainly it will still be here after everything else is gone, an endur-ing monument holding the 102-ton legacies of millions of Angele-nos. Landfills, Big Mike likes to say, are forever.

For now, Puente Hills is a living, breathing landfill—with a deadly "breath" expelled in massive burps that must constantly be siphoned off or risk disaster, a reeking, highly explosive, climate-destroying exhalation capable of turning green grass brown in short order. This property of buried garbage proved a difficult lesson in the bad old days of trash disposal early in the twentieth century, when cities routinely used trash and ash to fill in swamps and mud-

flats. (Such areas were regarded as bothersome wastelands imped-
ing progress back then; we call them irreplaceable, vital wetlands
and endangered habitats now that we've destroyed most of them.)
Housing projects, stadiums, parks and other developments that
were planted atop early fills suffered from unexplained stenches,
vermin infestations, swarms of roaches and, once decomposition
had reached critical mass, methane fires and explosions. Long Island, San Francisco and a hundred other places in between all
learned this the hard way: Trash can be deadly when you bury it.
Puente Hills's deep, aging refuse pile produces a constant flow of
31,000 cubic feet a minute of landfill gas (roughly half methane, half
carbon dioxide, with traces of various pollutants mixed in). If allowed to bottle up within the landfill, it could turn Garbage Mountain into something resembling a fiery trash volcano. This is the
flow that generates 50 megawatts of electricity around the clock and
provides power for all landfill operations to boot. You'd have to
cover 250 acres (or two and a half Disneylands) of sun-drenched
Mojave Desert with parabolic mirrors to generate an equivalent
output of solar power. At Puente Hills, the gas is expected to continue to flow for at least another twenty years *after* the landfill accepts its last piece of garbage.

Which is another way of saying that Puente Hills is big. Really
big. It covers 1,365 acres, half of that space devoted to buffer zone
and (oddly enough) wildlife preserve. The other half—a plot about
the size of New York City's Central Park—is devoted strictly to trash,
which by 2011 had reached heights greater than five hundred feet
above the original ground level. If the trash mounds of Puente Hills
were a high-rise, they would be among the twenty tallest skyscrap-
ers in Los Angeles, beating out the MGM Tower, Fox Plaza and Los

Angeles City Hall. Puente Hills is big enough to have its own microclimate and wind patterns, which the crews are constantly battling with berms and deodorizers and giant fans, trying to keep noxious odors from wafting through the surrounding bedroom communities of Whittier and Hacienda Heights.

Landfills are usually thought of, when they are thought of at all, as out-of-the-way things. Nobody really wants to think about what they contain: Puente Hills harbors millions of tons of moldering old carpet, even more rotting food and a good 3 million tons of dirty disposable diapers—2.5 percent of the total landfill weight consists of soiled Pampers, Huggies and all the other sweet names for some very noxious refuse. The material that seeps out of it, a noxious brew called "leachate," is so toxic that it has to be contained by multiple clay, plastic and concrete barriers, drainage systems and a network of testing wells just to keep it dammed and prevent it from poisoning groundwater supplies. The landfill workers didn't start trying to restrain this toxic goop by putting down strata of waterproof plastic liners under incoming trash until 1988—almost no American landfill did. So there are millions of tons of garbage at the bottom, oozing downward, a closely monitored potential time bomb. Every landfill started before 1991, when tougher federal regulations finally kicked in to make liners a requirement, is the same.

Yet this Garbage Mountain is not set in the hinterlands, neither out of sight nor out of mind. It lies smack in the middle of the most populous urban sprawl in America, the Los Angeles Basin, rising up to dominate its low-slung skyline for miles, a misshapen mound planted with thirty different species of trees and shrubs in a bold and ultimately futile attempt to mask its true nature.

THE STATE OF GARBAGE IN AMERICA: WHERE DOES IT GO?

By region, from least to most landfilling

Region	Landfills	Recycling/ Composting	Waste-to- Energy
New England	31%	29%	39%
West	52%	46%	2%
Mid-Atlantic	59%	27%	14%
USA Total	69%	24%	7%
Midwest	78%	22%	< 1%
South	79%	13%	8%
Great Lakes	81%	14%	4%
Rocky Mountains	88%	11%	1%

Source: Columbia University/BioCycle[1] Data from 2008, reported in 2010.
Percentages may not add up to 100% due to rounding.

PUENTE HILLS has been a trash destination for Los Angeles since the 1950s, back when it was an ordinary town dump—a rather small, scruffy one on the edge of a dairy farm. It wasn't until 1983 that it would be christened as the future of trash in Los Angeles, a model facility and the solution to what was then declared by political leaders and press headlines to be a "garbage crisis."

The garbage crisis turns out to be something that has been declared with surprising regularity throughout human history. It usually involves the question: *Where are we going to put all the trash?* After a number of false starts and grandiose promises, Los Angeles's leaders answered this question in 1983 by deciding (without actually saying so publicly): Let's bury it all in a canyon on the edge of the San Gabriel Valley and slowly turn it into a garbage mountain. Among other things, this decision guaranteed that, in thirty years, when the canyon was full and the landfill's state permit had

THE STATE OF GARBAGE IN THE WORLD: WHERE DOES IT GO?

By country, from least to most landfilling

Country	Landfills	Recycling/ Composting	Incineration
Germany	0%	66%	34%
Netherlands	1%	60%	39%
Austria	1%	70%	29%
Sweden	2%	49%	49%
Belgium	4%	60%	36%
Denmark	4%	48%	48%
France	32%	34%	34%
Italy	45%	43%	12%
Finland	46%	36%	18%
United Kingdom	48%	40%	11%
Spain	52%	39%	9%
Portugal	62%	20%	18%
USA	69%	24%	7%
Hungary	72%	18%	10%
Poland	78%	21%	1%
Lithuania	96%	4%	0%
Bulgaria	100%	0%	0%

Note: Incineration is for electricity generation and/or heating buildings.

Source: "The Sustainable Waste Management Ladder," Earth Engineering Center, Columbia University, based on Eurostat 2008 data.

Percentages may not add up to 100% due to rounding.

lapsed, another crisis would erupt, involving the same exact question: *Now where are we going to put all the trash?*

It would seem, then, that this is the wrong question to ask, at least if the goal is to permanently end the crisis rather than simply postpone a day of reckoning. After all, a landfill, by definition, will someday be full, and so all it does is enable the continued creation and flow of trash, rather than force a reconsideration of waste. A better question might be: *Why do we have so much trash, and what might we do to make less of it?* Eventually that question will have to be addressed somehow, as the cycle of crisis, finding a new place to hide trash, then returning to crisis cannot go on indefinitely. There simply are not enough affordable and convenient places for land-fills left in many parts of the country to continue repeating the cycle indefinitely. Certainly there are no other spots in Los Angeles to put another Puente Hills. That's part of the reason why the number of landfills and city dumps in America has dwindled from more than sixteen thousand authorized disposal sites in 1970 (along with ten times as many illegal dumps)[2] to just over 1,200 sanitary land-fills in 2011. The old sites are long since buried, capped over and, in several hundred instances, slated for extensive hazardous waste cleanup through the federal Superfund program.[3] The lack of local landfill space is also why Los Angeles is stuck with a plan conceived in boom times but untenable in a recession to use very expensive trains to transport trash two hundred miles into a neighboring county's deep desert beginning in October 2013. People are, unsur-prisingly, referring to this prospect as—you guessed it—a garbage crisis.

To put this in perspective, the very first documented trash crisis dates back 2,500 years to the ancient Greek capital of Athens, where city fathers grew alarmed by citizens' habit of hurling their refuse out of windows and doors where it clogged alleys, streets and walk-ways (a common practice in ancient cities and, until relatively re-cently, almost every modern city). The resulting hazard, filth, disease and odor led those forward-thinking Greeks (who by then had al-ready invented democracy, geometry, the Olympics, trial by jury, atomic theory and the Oedipal complex) to legislate into existence one of their most copied innovations, the municipal dump—the first in recorded history. They accompanied it with a law forbidding the disposal of garbage within one mile of the city limits, initiating the world's first anti-litter campaign.

Like so many other ancient advances, however, the Athenians' foray into the art of the landfill got mixed reviews (you try walking your trash a mile with only Bronze Age technology at your disposal). Soon the concept was lost and largely forgotten as the old practice of fouling the urban nest resumed from antiquity through the Middle Ages and beyond. The ancient Romans had some very advanced sewers and indoor toilets (using reclaimed bath water to provide the flush—another ancient first), which Pliny, author of the first encyclopedia, rated as ancient Rome's "most noteworthy" accomplishment. Archaeologists have found subterranean vaults in the ancient Roman city of Herculaneum (later buried along with nearby Pompeii by volcanic eruption) filled with ossified feces and small items of trash (pottery fragments, fish bones and other detritus from a pre-plastic world), suggesting the citizens of the Roman Empire used their advanced plumbing as something of a garbage disposal as well as a human waste repository. But this was not matched by any sort of house-to-house garbage collection or town dumps, and trash, primarily food waste and broken ceramics, accumulated in alleys and streets. Eventually, the layers of refuse became the foundation for stepping-stones, then new structures built atop the rubbish, gradually raising entire cities' elevations. (Bronze Age Troy provides a particularly dramatic ancient example of this process, as rubbish raised that city's average elevation at a rate of four to six feet per century.)

The town dump concept finally found official resurrection in the 1300s, when the French linked trash accumulation not to sanitation but to national security. It seems the mounds of stinking debris piled at the gates of Paris made it difficult for the city's defenders to spot approaching invaders, leading to a sensible new mandate to dump large items of trash farther off, keeping the gateways—and sight lines for spotting advancing enemy hordes—clear of debris. Unfortunately, this change in disposal rules did not extend to the Parisians' *tout à la rue*—all in the street—method of dumping everyday garbage in parts of the city well inside the city gates. The age-old practice had the unintended consequence of providing ample breeding grounds for the rats that carried the fleas that carried bubonic plague. Garbage, then, played a significant supporting role in the Black Death that decimated Europe not long after the great Parisian gate clearing.

Since then, the Old and New World trash policies have bounced from crisis to crisis, controversy to controversy—marked by battles over the practice of letting pigs roam city centers to devour garbage strewn on the streets, over dumping city trash in the ocean (the Supreme Court had to weigh in on that one in 1934 to save New Jersey beaches from floating New York refuse), over the smog generated by thousands of backyard trash incinerators in the Los Angeles Basin, over landfills that leaked awful odors into the air and toxins into water supplies, even over asking homeowners to separate their trash by type to aid recycling, a practice first instituted more than a century ago in New York City. Los Angeles mayor Sam Yorty won office in 1960 in large part by campaigning on a one-home, one-trash-bin platform that put an end to compulsory refuse and recyclables separation by homeowners, setting the cause of recycling back decades. There was the Virginia Garbage War of 1894, when irate residents of Alexandria began sinking trash barges on the Potomac to keep other people's refuse from entering town, and a century later, there were the Los Angeles Garbage Power Wars, in which community opposition scotched plans to replace old landfills with giant electrical generating stations that burned trash for fuel. Similar conflicts continue to brew today, as disagreements erupt over whether trash should be buried or burned; recycled at taxpayer expense or returned to manufacturers as their responsibility; shipped to distant landfills or accepted from distant countries (with Tennessee making its mark by welcoming, then incinerating, thousands of tons of radioactive waste from Germany and England). The same old question continues to animate the debate: *Where are we going to put the trash now?* Our entire elaborate waste collection, transportation and disposal system has for a century been built around this question, and the illusion that everyone's 102-ton legacy can be picked up piece by piece, week by week, and made to disappear, when in reality we have been building mountains with it.

EVERY MORNING, Monday through Saturday, a continual line of trash trucks come to Garbage Mountain to dump their loads, filing into the weighing station at the main gate, nine giant scales capable of measuring gargantuan loads (total weight minus the weight of an empty truck times $38 per ton equals the price of entry in 2011). Then there are the radiation detectors to stop the illegal dumping of radioactive waste (they're looking for low-level medical trash and similar materials, not terrorism-grade stuff). There are also random searches of trucks to screen for other hazardous and toxic materials, which are forbidden at Puente Hills, though they invariably sneak in—the alkaline batteries and half-full paint cans and aerosol bug spray cans that get slipped into garbage bins, and sometimes bigger and worse chemicals and substances.

Once through the gates, the scales and the screening, the trash trucks wind their way up to the foot of the new cell. Temporary roads for these garbage haulers are built of broken asphalt and other construction debris at the garbage mountain summit, where the active cell sits on a plateau of raw, naked dirt. The surface here is pounded and pulverized to a fine texture by the passing BOMAGs and dozers and other heavy vehicles. This dirt "skin" is a big part of why Puente Hills is not called a "dump" anymore, but is instead a "sanitary landfill." This technique was pioneered in the thirties at a dump in Fresno (now a national historic site) and perfected by the Army during World War II, though it was not widely adopted throughout the U.S. until the sixties and seventies. The dirt covers the trash buried beneath, sealing it in along with its aromas. "You're standing on top of five hundred feet of trash," one engineer observes, gesturing at the land, the enormous equipment, pretty much everything in sight, all of it atop a Grand Canyon filled with trash. His proud smile mirrors the expression mountain climbers often display after achieving some impossible summit. "Impressive, isn't it?"

Tanker trucks filled with recycled waste water drive in big, lazy orbits around the gritty plateau where the new cell is being built, spraying down the dirt to keep the dust clouds at bay, so the constant churn of big equipment doesn't send billowing brown clouds down into the neighborhoods below. Bulldozers erect a berm all around the cell to screen the sight and sounds from annoying the neighbors, too.

As the cell is filled, trenches are dug to accommodate lengths of wide plastic pipes that will be buried with the garbage and collect the landfill gas emitted as it rots. This in-trash plumbing is then connected to the web of pipes snaking across the surface of Garbage Mountain. Back in the eighties, the Puente Hills engineers decided to break with landfill tradition and stop merely "flaring" the gas—the practice of burning it inside a giant torch to keep the raw methane from entering the atmosphere, where it becomes a potent greenhouse gas—and instead put it to use for power generation. They soon ran into the same problem others had encountered when trying to mine energy from landfill gas: Over time, as the trash in the landfill decomposed and settled under its own weight, the pipes would crack, crush and break. The ingenious, low-tech solution— adopted first at Puente Hills, now employed all over the world—was to use plastic pipes of varying diameters and fit them together loosely, with plenty of overlap, like arms in a sleeve. As the trash mound settles, the pipe sections can move up and down at different rates and angles without damage, yet stay connected. Pumping stations create a slight vacuum inside this subterranean plumbing, which sucks the landfill gas in through those same loose joints and carries it to the surface, where the bristling network of pipes crisscrosses the surface of Garbage Mountain, resembling more than anything the big overhead heating ducts in open-ceilinged warehouse stores. Not only does this system provide an admirable fuel for electrical generation, it provides the added benefit of piping away a major source of landfill stink—the sulfuric smell of rotten eggs.

The smell close to the newly dumped raw, exposed garbage is another matter. It speaks of the moldering, sodden and swollen perfume of early rot—the amalgam of smells characteristic of any well-used trash can opened on a hot summer day, times a million or so. But mostly, there is surprisingly little of the scent of putrescence atop Garbage Mountain. This is because, each day, once the last load of trash has been dumped, clawed, pushed and crushed, the new cell is covered over with that six- to twelve-inch skin of clean dirt mined from a nearby "natural" hill, burying the stink with the trash, along with anything that might attract vermin or seagulls. Old-style dumps were rat havens, but the modern landfill operators are confident that they offer little hospitality for scavenging rodents, thanks to the crushing force of the BOMAG coupled with the dirt cap. This confidence was put to the test a few years back, when some enterprising landfill engineers fitted four rats with radio transmitters and released them into the fill. Three were done in within an hour—presumed crushed and buried; the fourth met its doom in less than a day.

Seagulls are another story: Their garbage-loving traits are the bane of landfill workers everywhere, as the birds amass wherever the smell and sight of human debris can be found. They are particularly despised for their habit of scooping up garbage in their beaks, flying off to get away from their fellow scavengers, then dueling in aerial battles during which some offensive piece of garbage is invariably dropped. All too often this little garbage bomb lands in a nearby residential area.

Homeowners are quick to call and complain when some odorous hunk of rotting meat ends up festooning their hydrangea beds or rose gardens or kid's swing set. Landfill workers must therefore take daily precautions against the marauding flocks of gulls—stringing fishing line from movable poles surrounding the cell each day, which disturbs the gulls' flight patterns. Seagulls instinctively glide to a landing in long approaches, which the fishing line seems to impede. The gulls' vision, more sensitive to ultraviolet light than that of humans, is believed to perceive the fine monofilament line as a disturbingly bright blue barrier, though it is almost invisible to human eyes.[4] Whatever the reason, the lines do seem to stop or at least slow the seagulls' access to the active cell and the prized garbage within. They perch nearby, though, eyeing the fresh, smelly trash balefully, squawking and flapping but for the most part foiled as Big Mike zooms by, putting the trash to rest. Sometimes, though, the gulls find a way in, and for those occasions, the landfill guardians have guns that fire noisy blanks and two remote-controlled airplanes that buzz the seagulls and drive them off (and yes, everybody wants a turn to fly the planes). But that's just a sideshow to the action at center stage, where Big Mike and his cohorts are constructing Los Angeles's mightiest, if unintended, monument.

The engineer at the summit was right: This mountain of trash is impressive. It's also compelling, revelatory and horrifying all at the same time, possessed of that frightful beauty-beast admixture that can arise when vast natural and industrial landscapes are forcibly grafted onto one another—the Tinkertoy gas pipes snaking through wildflower fields, the scents of fresh pine mulch and fresh garbage vying for supremacy. A road on the backside of the mountain winds along the artificial "benches" that are built at forty-foot intervals, tiers of garbage capped with clay, soil and greenery. The road leads to the trash power generators, the landfill gas pipe junctions, various other pieces of garbage infrastructure, the boundary with Puente Hills's unlikely neighbor, the sprawling Rose Hills Cemetery, and the active dump site topside. Because garbage settles as the years pass no matter how tightly it's mashed by the bite and crush of Big Mike's compactor, the roadbed settles, too, over time. The engineers can calculate this effect on average, but depending on whether the deep underbelly of a particular stretch of road is mostly crushed construction debris or mashed Barbie dolls or old carpeting—it's impossible to know which without digging it all up— the rate and distance of settling varies yard by yard, causing the road to undulate like a roller-coaster track. This provides a bit of a thrill for the riders on the guided bus tours at Puente Hills, which are almost always full at the landfill that bills itself as the Disneyland of dumps.

"There is no other place like it, and no other job like it, either," Big Mike says, gazing fondly at his dusty, noisy workplace. This observation is accompanied by a sigh of satisfaction tinged with regret, because soon, Big Mike knows, it will end. Soon the mountain will be finished, though not gone, of course—a landfill is never gone. It's the question of what's next that has not yet been resolved, that L.A. and the rest of the country are trying to puzzle out, and that will have lasting consequences no matter how it's answered: Is it time to dump the dump as the centerpiece of waste? Or time to hedge our bets once again and find even bigger dumps to take their place?

SELECTED PRODUCTS, PERCENTAGE BY WEIGHT
OF TOTAL LANDFILLED TRASH

Furniture & Furnishings	6.1%
Clothing & Footwear	4.9%
Wood Packaging	4.9%
Corrugated Boxes	3.2%
Disposable Diapers	2.4%
Beer & Soft Drink Bottles	2.3%
Bags, Sacks & Wraps	2.2%
Carpets & Rugs	2.0%
Rubber Tires	1.9%
Junk Mail	1.1%
PET Plastic Bottles	1.1%
Major Appliances	0.8%
Trash Bags	0.6%
Newspapers	0.6%

Source: EPA*

*Although the EPA data on the quantity of waste generation in the U.S. is flawed, its analysis of the vi composition of trash depicted here continues to be useful and reliable. These calculations are informed in part by studies of real-world samples of typical Americans' trash—how much of it is plastic, metal, paper, food scraps and so on. These figures are expressed in the EPA annual municipal solid waste reports as percentages of the total waste stream, as in the example of carpets and rugs, which are reported to comprise 2 percent of the total weight of trash sent to landfills. This is a different methodology from the flawed material flow analysis used to calculate total tonnage of waste. Extrapolating national estimates from real-world samples is a tried-and-true, scientifically valid technique.

What's in Our Trash?

Percentage of materials we throw away, by weight,
before recycling and composting

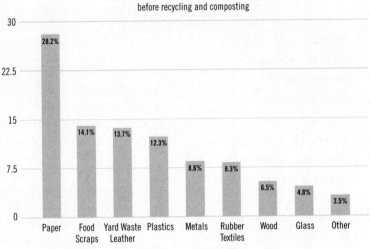

28.2%	14.1%	13.7%	12.3%	8.6%	8.3%	6.5%	4.8%	3.5%
Paper	Food Scraps	Yard Waste Leather	Plastics	Metals	Rubber Textiles	Wood	Glass	Other

What's in Our Landfills?

Percentage of product categories we bury in landfills, by weight,
after recycling and composting

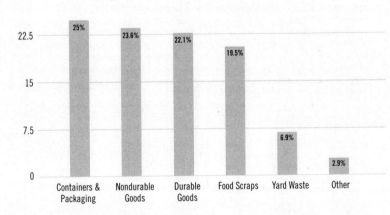

25%	23.6%	22.1%	19.5%	6.9%	2.9%
Containers & Packaging	Nondurable Goods	Durable Goods	Food Scraps	Yard Waste	Other

PIGGERIES AND BURN PILES: AN AMERICAN TRASH GENESIS

IMAGINE A CITY SO FULL OF GARBAGE, MUCK, HORSE manure and standing sewage that sailors can smell it six miles out at sea. Where a proper gentleman has to toss a coin or two to a broom-wielding street urchin to sweep a path through the knee-high debris just so he can get in his front door. And where pigs trundle down sidewalks and dodge traffic, rooting through the garbage that the locals simply throw out their doors and windows, in vain hope that the ravenous animals will clean up some of the mess.

Welcome to New York City at the dawn of the twentieth century. Welcome to American urban life before the rise of the landfill. Even

Charles Dickens, no stranger to the foul side of city life, was horrified by it: "Take care of the pigs," he wrote in *American Notes*. "Ugly brutes they are; having, for the most part, scanty, brown backs, like the lids of old horse-hair trunks: spotted with unwholesome black blotches . . . They are the city scavengers, the pigs."

The winding road that led us to flawed garbage numbers in Washington and a garbage mountain in Los Angeles begins 120 years ago in the New York of the 1890s. Widely considered at the time to be among the most exciting, vibrant and corrupt cities in the world, it was also the foulest, smelliest and dirtiest—truly a city of garbage. This is arguably the fertile soil in which America's modern waste-management system begins.

At the opposite end of that history stands Puente Hills, the literal and figurative pinnacle of that same waste-management system, representing both its greatest success (handling huge amounts of waste) and greatest failure (enabling huge amounts of waste). For decades it has been a model for other dumps both public and private worldwide. The mega-landfill's birth and evolution, then, provide a veritable history lesson on trash in America, which makes it a five-hundred-foot-tall 3-D chronicle of the modern consumer economy as well, the back end of which can be no better represented than by a mountain of garbage.

The modern landfill's roots can be traced back to a City Hall regime that initially refused to use them: the corrupt Tammany Hall administration of the nineteenth and early twentieth centuries, a political machine that ran New York City like a feudal empire. This isn't just metaphor. At the time, deaths in New York City from preventable diseases such as cholera matched those of Europe in the Dark Ages. The mixing of trash and disease-bearing human waste in the streets and gutters, coupled with an utter disregard for safe drinking-water supplies, made it quite certain that the rotting mess that paved the Big Apple would also breed one epidemic after another. America's center for high culture had also mutated into its ultimate bacterial culture.

What garbage collection the City Hall bosses saw fit to make found its way onto barges of refuse dumped into the ocean, where it soon became the Jersey Shore's problem. This was billed as an improvement for New Yorkers over the earlier practice, discontinued in 1872, of simply dumping garbage from a wooden platform over the East River. Garbage was strewn everywhere on the busy streets of the New York boroughs—food waste, ashes, human waste and considerable amounts of animal droppings, primarily from the 120,000 horses plying the roads in those pre-automotive times. Every day those horses deposited on the order of 1,200 tons of manure and 60,000 gallons of urine on the cobbles and asphalt of New York. Thousands of horse carcasses a year were just left where the overworked animals fell (fifteen thousand in 1880 alone). The wandering herds of pigs allowed to forage through the mess in lieu of regular street cleaning did consume some of the garbage, but they also added their own droppings to the overall muck. Cleanup, to say the least, was spotty; in a city where the population was doubling every decade for most of the nineteenth century, waste had become public enemy number one.

New York was not unique in this regard: Chicago, Philadelphia, Washington, D.C., and other cities large and small were besieged by waste and ineffectual attempts to collect, dump or burn it. More than two hundred towns with populations over ten thousand built piggeries where raw garbage served as the feed, as what passed for waste experts at the time estimated that seventy-five pigs could dispose of a ton of garbage a day—and provide revenue and meat at the same time. New England led the nation in pursuing this waste-to-swine strategy; turn-of-the-century New Haven sent all its wet garbage, 5,400 tons of it a year, to pig farms, while Worcester, Massachusetts, proudly kept two thousand garbage-swilling swine at its forty-acre piggery near the city limits.[1] A network of private garbage haulers in New York City cut deals with major hotels and restaurants to convert their food scraps into food for swine, which was then sold to local farmers. The piggeries provided more of a boon to garbage entrepreneurs and pork suppliers, along with the avaricious politicians who sometimes favored them, than a reliable means of stemming the tide of trash, and of course did nothing for nonedible refuse. Still, the practice, with the added innovation of cooking the garbage as a means of killing pathogens, persisted until the 1960s, when evidence that it could spread disease to both swine and humans became impossible to ignore.

Even the White House was plagued by the smells of festering garbage back then, not to mention the rats and roaches that nested in it. Presidents traditionally had to hire their own private trash haulers until well into the nineteenth century just to keep the White House garbage-free and the vermin under control.

But it was the legendary filth of New York City that provided the tipping point for the next evolution in garbage. Voter outrage over trash, corruption and endless scandal finally led to the ouster of the Tammany political machine in the elections of 1894. Soon after taking office, the new reformist mayor, William Strong, sought a superstar to accept an unprecedented mission: to bring the city back from trash Armageddon.

The mayor's first choice, Teddy Roosevelt, turned down the job, taking the post of New York police commissioner instead. He deduced that this position would be a better springboard to higher political office for the future president than garbage czar, letting him build a name for ferreting out corruption rather than manure. The job of cleaning up New York then fell to Colonel George E. Waring, a Civil War veteran who, before his military service, had worked as the city engineer responsible for reclaiming the swampland that would become New York's Central Park. Waring had supervised the design of a drainage system that created the park's famously scenic lakes and ponds while leaving the rest of it dry. He had gone on to engineer an affordable and efficient dual sewer and drainage system for Memphis that kept storm runoff and septic waste separate. This protected the city water supply from contamination, ending almost overnight the cholera and other waterborne epidemics that had beset "The River City" for decades. Reforming New York's sanitation department seemed a natural fit for this leading sanitation engineer of the day, who harrumphed into office asserting that he wished to be called "Colonel," not "Commissioner," throughout his tenure. His workers were required to salute.

Waring's first move certainly drew more on his military training than engineering skill: He amassed an army, with broomsticks and ash cans as their weapons. Waring converted the desultory ranks of street cleaners previously known for collecting more bribes than garbage into a force of two thousand sanitation soldiers who marched in formation, passed by in lockstep for review by city officials, and were clad completely in white, from pants to pith helmets. New Yorkers stood openmouthed in shock when this new legion of street cleaners first appeared.

Initially mocked, within a year these "White Wings," as Waring's troops were nicknamed, had become icons, their white garb soon an international symbol (and eventually a cartoon cliché) for street cleaners everywhere. Their habit of menacing and occasionally roughing up litterbugs became legendary. They paraded before cheering crowds who marveled at the transformation of whole swaths of the city, the streets clear and the smells quashed—not just in upscale neighborhoods, but in the poorest parts of town as well. The White Wing treatment even extended to the gang-dominated and crime-infested tenements of Five Points, where gangster Al Capone got his start, where the notorious Tombs city prison long stood, and where the Civic Center government complex now resides. The muckraker Jacob Riis, whose investigative reporting on unsafe drinking water has been credited with saving New York from cholera outbreaks and whose impassioned writing about life in the city's slums sparked the creation of a system of neighborhood parks, wrote: "It was Colonel Waring's broom that first let light into the slum . . . His broom saved more lives in the crowded tenements than a squad of doctors." The *New York Times* reported after seven months of Waring's sanitation leadership: "Clean streets at last . . . Marvels have been done." The paper later would eulogize him by asserting: "There is not a man or a woman or a child in New York who does not owe him gratitude for making New York in every part so much more fit to live than it was when he undertook the cleaning of the streets." It's safe to say that no sanitation official before or since has ever been so celebrated. And none have influenced his successors and peers so profoundly, for his model is essentially the one in place now in America.

Waring's attack on garbage began with street cleaning coupled with a new form of garbage collection in which recycling, "reduction" and recovery were emphasized as never before. Waring's price for regular and bribe-free trash collection and street sweeping was a requirement that households divide their waste into three bins: ash (from burning coal), food waste, and all other rubbish, which had to be bundled by type. Waring wanted to build a system to recycle useful materials such as scrap metal, rags and paper; incinerate burnable garbage; and reduce (or render) organic waste, including animal carcasses (remember those fifteen thousand dead horses?) into "garbage grease." The product of the rather noxious rendering plants was then sold for soap, perfumes and lubricants, along with an incineration byproduct referred to as "residuum" that was sold as a cheap fertilizer. The principles of reuse and recycling Waring pursued were not new—various industries, such as printing and paper making, had recycled since before the American Revolution, and trash scavenging had long been a source of work and income, embraced by waves of immigrants as one of the few jobs never denied them over issues of language, race or culture. Recycling in the broad sense appears to be one of the most ancient human instincts, dating back to our pre-human, tool-making hominid ancestors, whose ancient stone implements reveal that they did not throw away a broken flint knife, but instead recycled it, shaping it into some other, smaller tool or weapon. But Waring took recycling to a whole new level in America's fastest-growing metropolis: He institutionalized and regulated the idea. He set an entire city on course for the kind of sustainable handling of waste that America has been trying in fits and starts to perfect ever since.

His methods included cooking the garbage with steam or naphtha, pressing out the liquid into settling tanks and skimming the grease off the top. In this way, a ton of "summer garbage" (minus the coal ash and other nonrecyclable trash) broke down into forty pounds of garbage grease (2 percent of that ton of waste) and four hundred pounds (20 percent) of "tankage," which was the stuff that settled and could be used as fertilizer, as it contained ammonia, phosphoric acid and potash. Seventy-one percent of that ton of garbage—1,420 pounds—consisted of water cooked off during the process. From the leftover 7 percent—the "rubbish"—lead from tin can lids was melted and reused for solder, and the remaining tin was pressed into blocks for making window counterweights. Waring reported that one ton of municipal garbage processed in this way could generate $2.47 of revenue from these recovered materials— which would be a respectable $64 in 2010 dollars.

Once the reusable materials had been stripped out of the trash flow, whatever was left would be taken to dumps outside the city. They were not the sealed and antiseptic Puente Hills–style of contemporary American landfills. They were traditional, stinking open dumps prowled by scavengers and infested with rats and insects. Still, they were a vast improvement over using the streets or the ocean as a dump. Their existence helped establish a new principle that quickly became dogma: One of local government's basic functions was to keep the streets clean by removing refuse to a safe distance for proper, well-thought-out disposal. Waring argued that civilization and human health depended on this basic service, and he became the toast of the town as a result, having succeeded in bringing New York City back to a point that the ancient Greeks had reached five centuries before the birth of Christianity. Which is more than anyone else had accomplished before him.

STILL, CELEBRATED and admired as he and his White Wings were, parts of Waring's plan didn't go over so well. He had badly overestimated the public's willingness to participate in the reinvention

CARD OF INSTRUCTION FOR HOUSEHOLDERS

Put into Garbage Receptacles

Kitchen or Table Waste, Vegetables, Meats, Fish, Bones, Fat.

Put into Ash Receptacles

Ashes, Sawdust, Floor and Street Sweepings, Broken Glass, Broken Crockery, *Oyster and Clam Shells, Tin Cans.

Put into Rubbish Bundles†

Bottles, Paper, Pasteboard, etc. Rags, Mattresses, Old Clothes, Old Shoes, Leather Scrap, Carpets, Tobacco Stems, Straw and Excelsior (from households only).

*Note. Where there is a quantity of shells, as at a restaurant, they must be hauled to the dump by the owner.

†All rubbish such as described in this third column must be securely bundled and tied, or it will not be removed.

Reverse of card

It is forbidden by city ordinance to throw any discarded scrap or article into the street, or paper, newspapers, etc., ashes, dirt, garbage, banana skins, orange peel, and the like. The Sanitary Code requires householders and occupants to provide separate receptacles for ashes and garbage, and forbids mixing these in the same receptacle. This law will be strictly enforced.

of waste. New Yorkers, Waring complained, were too "obdurate" to follow his trash separation rules. When it became clear rigid enforcement—Waring's preference—wasn't practical, he instead built a sorting facility of his own, where useful materials could be pulled out of the garbage after collection but before disposal. This innovation was a forerunner of what is now an industry standard for sorting trash, the Materials Recovery Facility (which people conversant in waste-speak simply call the MRF, pronounced "the Murph"). Puente Hills, for instance, has a state-of-the-art Murph. But the concept was pioneered more than a century ago by Colonel Waring. Unfortunately, it was a costly and dangerous process then, and without Waring's ironfisted supervision, it soon fell by the wayside.

For there was an unintended consequence of Waring's success: New York experienced the retreat of civic outrage and, all too soon, a return in the very next turn of the election cycle of the old Tammany regime. The streets were clean and the scoundrels were voted back in.

Waring's appointment ended with the ouster of the mayor who had given him command over New York's trash. He departed his White Wing militia and agreed to lead the construction of a sewer system for Havana, Cuba, where waterborne diseases were then rampant. While completing that task, he contracted a fatal case of yellow fever and died soon after returning home.

Those who predicted the end of his tenure would mark a resumption of the bad old days turned out to be mistaken, at least in part. Even without his doleful, demanding, ramrod military presence, New York City streets never returned to their old depths of squalor. Even the most venal ward heelers had come to understand that they could not survive long in office should there be a return to the knee-deep "corporation pudding," as New Yorkers derisively termed the muck on the streets that the White Wings had cleared away.

So Waring's successors stuck with the most visible of the colonel's ideas. They even pioneered a major innovation: the first waste-by-rail project. Electric trolleys were enlisted to streamline the disposal of the massive amounts of coal and wood ash generated daily by the city's residents and factories, a stream to which all sorts of other garbage soon were added. A growing city found a new use for this waste: Hauled to a massive ash dump on Barren Island in Jamaica Bay, the refuse was then used to fill in New York marshlands that were subsequently covered over to build houses and marinas. Once that effort was complete and no more waste could be squeezed onto Barren Island, a new ash repository opened—the Corona Dump in Queens, where the massive amounts of toxin-laden ash and assorted other trash were again used as landfill to reclaim salt marshes for development. When that was done, the black and gray waste accumulated into a smoldering mountain of ash ninety feet high, a fetid, volcanic landscape. The place was depicted by F. Scott Fitzgerald as "the Valley of Ashes" in *The Great Gatsby*. In one of the most celebrated literary commentaries on waste and the unintended consequences of American industrialization, Fitzgerald called the dump:

> a fantastic farm where ashes grow like wheat into ridges and hills and grotesque gardens; where ashes take the form of houses and chimneys and rising smoke and, finally, with a transcendent effort, of men who move dimly and already crumbling through the powdery air.

In an early demonstration of the potential value and longevity of even the most repulsive landfills, the Valley of Ashes and "ash mountain," as the looming peak of cinders and garbage had been nicknamed, became a golf course and country club with the dump masked by the thinnest of disguises, the ash heap literally looming over the fairways. Only the politically wired Tammany Hall business kingpin who owned the dump could have managed such a juxtaposition, especially given that the land was so contaminated it had to be doused daily with disinfectants for the protection of the golfers. The city eventually bought the dump property in the 1930s, and workers leveled the golf course and the entire cinder mountain. Then they covered over the wasteland and planted it as a park. In short order the dump site became the scene of the celebration of the 150th anniversary of George Washington's presidential inauguration—a little affair better known as the World's Fair of 1939.

The enormous Corona Dump by then had been rebranded Flushing Meadows. Over two years, more than 44 million visitors flocked to Flushing Meadows to view "the world of tomorrow" on display at the fair where, among other things, they witnessed the burial of a time capsule holding the writings of Albert Einstein and Thomas Mann, a Mickey Mouse watch, a selection of seeds and a pack of Camel cigarettes. It was sealed for five thousand years and buried fifty feet below ground into the residue of the Corona Dump.

The site also hosted the 1964 World's Fair a quarter century later, with its iconic, twelve-story Unisphere statue of the earth still marking the spot. The National Tennis Center, where the U.S. Open grand slam tournament is held annually, is now located atop the former dump, as is a park, a zoo, two lakes (one for sailing), a golf course, the Queens Museum of Art, the New York Hall of Science, the now-demolished home of the Mets baseball team, Shea Stadium, and its successor, Citi Field. The Valley of Ashes may still lurk deep below—landfills really are forever—and the park's twin artificial lakes are plagued with pollutants, but few who walk the pleasant grasses of Flushing Meadows or watch the Mets in their state-of-the-art ballpark have a clue.

Although the ash had a place and a purpose in the aftermath of Waring's reign, the ocean dumping he had tried to end resumed for other types of garbage for another three decades. That was the end of his early reuse-repurpose-recycle experiment. Compared to Waring's painstaking sorting and recycling process, using the ocean as a trash bin was cheap and easy, especially when the prevailing tides made it New Jersey's problem—at least until the U.S. Supreme Court held New York City liable for damages and forbade the practice in the 1930s. (This ruling applied only to the "public nuisance" of residential garbage dumped in the ocean. In a decision that was perfectly logical legally and yet makes absolutely no sense outside the courthouse, the Supreme Court exempted businesses and industry from its ruling, giving them a free pass on marine dumping and pollution for decades longer.)

Still, Waring's influence extended beyond both the boundaries of New York and his time in office, and his methods were widely copied. Within a decade, a majority of American cities had created sanitation departments modeled on New York's, an enduring reform. Others were not so long-lasting. More than 180 incinerators were built in the belief they were the quickest fix for dealing with trash. But most were shut down within a decade because of unexpectedly high costs, overhyped yet under-performing technology and horrendous pollution. The reduction plants where garbage was stewed to produce grease and fertilizer remained in vogue a bit longer than incinerators, but they were so noxious, contaminating air and water supplies, that they, too, were soon abandoned, with nearly all of them shut down by the time of the Great Depression. In many ways, waste management returned to ground zero: It was all about finding places to put the trash.

But not everywhere: Some of Waring's ideas struck a potent chord in Los Angeles, which became more enamored of trash incineration than almost any other city in America. In a region where most every home had a backyard—unlike the dense streets and tenements of old New York and other East Coast cities—backyard incinerators soon were considered a veritable birthright and a necessity. Businesses and factories began burning their trash as well. What couldn't be burned was hauled by private trash collectors and scavengers to open dumps scattered around the Los Angeles area, where burning was also used to keep the trash volume down.

The predictable result of making bonfires out of trash, given the characteristic wind patterns of the Los Angeles Basin that trap and concentrate fumes and smoke of all kinds, sparked the dawn of the age of smog. It's been largely forgotten, but L.A.'s worst-in-the-nation air quality woes long preceded California's famous car culture. By 1903, the choking haze had gotten so bad that citizens woke up one day convinced that they were experiencing an eclipse of the sun. Decades later, the rise of the automobile multiplied the incipient smog problem. Then the air grew markedly worse during World War II, when defense plants, naval yards, the aviation industry and other heavy manufacturing expanded rapidly in the Los Angeles area as part of the war effort and the postwar boom times, contributing their own factory emissions along with greater amounts of waste to be burned. In the space of five years, Los Angles went from the ninth largest American city to third on the list, and many of the newcomers happily joined the ranks of the backyard trash burners. Yet the connection between the poor air quality and public infatuation with setting fire to garbage never seemed to sink in. It was the factories that bore the brunt of blame on July 26, 1943, Los Angeles's infamous "Black Monday," when noxious fumes engulfed downtown, choking pedestrians and reducing visibility to three blocks. But when a reviled chemical plant thought to be responsible halted

BACKYARD TRASH BURNING FACTS

- Emits dangerous levels of dioxins, poisonous and cancer-causing compounds formed by low-temperature, smoky, inefficient combustion. Dioxins have no human uses or value, and are not made intentionally.
- Emits soot and fine particulates, which can cause emphysema and lung cancer and aggravate asthma and other lung ailments.
- Emits polycyclic aromatic hydrocarbons (PAHs), which can be highly toxic, have been linked to cancer and can cause birth defects or fetal death.
- Twenty families burning trash in their backyard put out more cancer-causing dioxins than a modern industrial facility that burns the trash of 150,000 families.
- In 1987, industry accounted for about four-fifths of total dioxin emissions in the U.S. By 2010, dioxin emissions were reduced by a factor of thirteen because of industrial pollution controls. But of the dioxin emissions that remained, home trash burning was responsible for nearly two-thirds of them.

operations, nothing changed. The smog and stench continued. For decades.

New York City had an analogous crisis in this same era, as incineration made a big comeback there after World War II as well. At its height in the fifties and sixties, New York boasted more than seventeen thousand apartment-building incinerators and twenty-two big municipal incinerators gobbling up a third of New York's waste—and casting a pall of smoke and soot throughout the city. The city council banned most of the residential incinerators in 1971, and public pressure soon shut down the municipal plants as well in favor of landfilling, first at Fresh Kills landfill on Staten Island, then out of state.

Several studies in 1950 by Los Angeles County health officials sensibly and correctly attributed the growth of L.A. smog to multiple sources, with trash pinpointed as a major culprit: backyard incinerators, open burning of waste at dumps and incineration of sawmill waste. Following votes by county supervisors and the state legislature, delayed but ultimately undeterred by lawsuits from business interests and chambers of commerce who asserted a constitutional right to pollute air and water, the first local pollution control district in the nation opened for business in 1950. All smokestack facilities in the Los Angeles area were regulated with pollution permits and gradually tightening emission limits. The new pollution police were able to ban open burning at dumps, reduce heavy smoke plumes from factories and order oil refineries to cut noxious sulfur dioxide emissions long before federal officials acted (or had the power to act) on air quality. Next, gas pumps were refitted with the now common hose sleeves that keep damaging gasoline vapors from polluting the air. The local pollution control districts eventually evolved into the regional air quality management districts in place in California today—a system that pioneered most major advances in air pollution control that are now nationally ubiquitous, from cleaner fuels to catalytic converters to modern electric cars.

But when it came to the three hundred thousand to half million backyard incinerators in Los Angeles, the so-called Smokey Joes, and the estimated 500 tons of sooty, toxic air pollutants they churned out each day, the story was different. People simply did not want to give them up.

For decades, Angelenos had been encouraged and expected to incinerate their trash. It wasn't just convenient, they had been told, it was a civic duty, a way of avoiding costly trash collection services, taxes and bills. They did not want to give up their burn barrels and piles of smoking rubbish, and they let their elected officials know it. The thinking was that a bit of soot and smoke in the backyard couldn't possibly be as big a problem as massive refineries, factories and freeways full of cars. Besides, where would all the trash go? What would it cost? Who would pay? Public sentiment clearly tilted in favor of letting officials address the perceived bigger smog culprits first, the smokestacks and the tailpipes, while leaving those deceptively small backyard burners alone. And so the incinerators would not go gently.

Neither did the million filthy, soot-covered smudge pots that orange growers and other farmers employed to protect crops from freezing overnight, burning old crankcase oil, discarded tires— anything cheap and combustible, which always meant dirty and toxic. It took years to convince them to switch to cleaner devices and

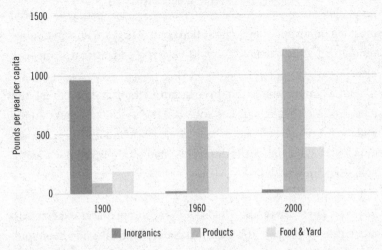

Waste Evolution 1900–2000

Pounds per year per capita

Inorganics Products Food & Yard

fuels, due to the prevailing and utterly false belief that the smoke "helped hold the heat" close to the ground. Incinerator manufacturers fought back to protect their interests, too, at one point trying to make their products appear more cuddly by marketing sheet metal incinerators shaped like little houses and featuring smiling faces painted beneath stovepipe chimney "hats."

It took seven years of failed attempts to finally pass the ordinances to ban incinerators countywide in 1957. The smog had grown so bad by then that it became nearly impossible to dry clothes successfully on outdoor laundry lines without them absorbing a rain of black soot. Complaints about the dirty byproducts of backyard burning finally matched the defenders, and politicians felt sufficiently safe to act: no more burn barrels, no more happy-face incinerators. Jail and a five-hundred-dollar fine awaited illicit burners, and Smokey Joe was finally toast.

As predicted, the home incinerator ban led to greater volumes of trash in need of disposal, which meant new trash hauling services by both government and industry arose to meet that need. And the garbage had to go somewhere once it was picked up, too. A web of dumps ringing the basin soon opened to accommodate the new and rapidly growing river of trash—growing because Los Angeles was growing, with bean fields and orange groves converted on a daily basis to postwar, GI Bill–financed suburban housing tracts. Along with the real estate boom, trash became a growth business as never before.

One place in particular drew Los Angeles's new mounds of garbage—the area surrounding the eastern L.A. County foothills bordering the San Gabriel Valley. Strategically located in an area with ample open land, it straddled a confluence of major highways and lay near several populous communities—the ideal mix of convenience and seclusion for trash disposal. There was not a single large repository for garbage opened then, but a profusion of small, privately owned garbage destinations. They were open dumps, like the vast majority in America at the time, where refuse was tossed and piled and, in many locations, burned. In short order this area of L.A. bore a nickname worthy of Fitzgerald: the Valley of the Dumps.

Demand for dump space began mounting then, not only in Los Angeles, but also nationwide. Bans on incinerator and backyard burn piles were only part of the reason. Another trash multiplier had arrived right around the same time: the rise of America's new consumer culture and the disposable economy ushered in with it. A new tidal wave of trash began to crest then, combining the old refuse that once had been burned with a new flow of disposable trash, containers and short-lived products never before seen. Consumption and garbage became more firmly linked than at any other time in history, with the disposal of products and their packaging displacing other categories of household waste for the first time in our trash history. That trend hasn't changed since 1960.

The age of the plastic bag was upon us.

FROM TRASH TV TO
LANDFILL RODEOS

Now and then when the wind blows just
right and the day's garbage is still baking uncovered in the Southern California sun, a flock of strange birds can be seen wheeling
above the Puente Hills landfill. Upon closer inspection, however,
some of these fliers turn out not to be birds, but escaped plastic
grocery bags, which are woven like veins throughout every load of
trash dumped in every city landfill in America. Sometimes a few
break free of the piles of dirty napkins, spent kitty litter and broken
glass holding them down, and they scuttle like urban tumbleweeds
across the jagged top of the trash cell, then take flight. This is one

of the myriad ways plastic trash makes its way into streets and rivers and oceans, and demonstrates the drawback of engineering products with useful lives that last the half hour or so it takes to bring groceries home from the market, but which possess a second life as refuse that can last a thousand years or more.

Flocks of flying immortal bags are a signature element and a unique hallmark of the disposable age of plastic—a trash challenge Waring and his White Wings never had to deal with or imagine. And yet, despite that, and despite the noise and scale and stench (which really isn't all that bad, except on the days the sanitation engineers "sweeten" the fill with sewage sludge to jump-start methane production), there is a weirdly beautiful aspect to this place, even to the strange plastic flock flapping and twisting above. To perceive this side of the landfill, one need go no farther than the expression on Big Mike Speiser's face when he considers his workplace. "We accomplish something here every day," he says. "It can be strange, it can be loud, but we're proud of this place, proud of what we do. There is a kind of beauty here, or there will be someday." He gestures at the oak trees planted in the distance on older sections of Garbage Mountain. "Someday all this will be a park."

It's clear the BOMAG master takes pride in Puente Hills's reputation for being a well-run Disneyland of dumps. Even the neighbors who complain about smells and dust and toxic leaks concede that much (which isn't to say they share Big Mike's love of the place—they can't wait for it to shut down). But Big Mike revels in being a landfill ambassador, talking with the press, demonstrating his driving skills for a *National Geographic* film crew, chatting with tour groups. He has represented Puente Hills in years past at the national trash Olympics, a not-quite-yearly event sponsored by the Solid Waste Association of North America. This is a gathering of heavy equipment operators from the nation's waste-management departments and companies, who compete in a series of Olympic-style events set up at a host landfill. At this combination rodeo and monster-truck rally, the contestants must perform various feats of "trash-tacular" skill with their big machines: pass through orange cones with only four inches of clearance on either side, navigate an obstacle course, do some precision blade drops and push a load of gravel to an exact location without spilling. Big Mike has several gold medals to his credit. His colleagues have to point this out, as he won't; he just nods shyly in acknowledgment once the subject is disclosed.

After the competition, the large men of the landfill fraternity— most of them seem to share Big Mike's mountainous physique— gather to cool off over a cool drink and to swap stories about the weird things that inevitably turn up at landfills. There's the mounted deer heads with their glassy eyes and broken antlers, the yellowed dentures scattered in the debris, the occasional bowling ball, the coffin (an empty discard), the mannequins (those always give the dump workers a start, arms or heads poking up out of the cans and Hefty bags). And there are the real bodies, too—because what better place could there be to dispose of the evidence of a serious crime? Puente Hills has had its share: the skull found in the suit-case, the body rolled up in a rug, the rumors of ritual crimes in adjacent Turnbull Canyon, the bloody Santeria shrine found nearby. Then there was Robert Glenn Bennett, who disappeared February 16, 1983, from his maintenance supervisor job at the sanitation district's water reclamation plant next to the landfill. The fifty-one-year-old never showed up for a part-time bartending gig that Wednesday night after he finished at the plant; witnesses said he argued earlier with a gardener on his staff, who was angry at being denied a promotion. The gardener had a record of minor drug, theft and vandalism offenses, of threatening fellow employees and of exposing himself on the job. Blood matching Bennett's type was found in the parking lot and on a dump truck, and the gardener, John Alcantara, became the immediate prime suspect. But despite an extensive and unpleasant search of mounds of garbage at the landfill, which police detectives figured to be a likely place for Alcantara to have stashed both body and murder weapon, no evidence could be found linking him to Bennett's disappearance (or, for that matter, to prove conclusively that Bennett was dead). Twenty-five years would pass before the case would be solved and Alcantara would be convicted of the murder. A witness finally came forward and told police that Alcantara confessed to shooting Bennett in the head, dismembering his body, and then, as the police always suspected, the body had been buried at the adjacent landfill. That's one of the darker legacies of Puente Hills: What's left of Robert Bennett remains there to this day, hidden beneath thousands of tons of trash in the Main Canyon section of the landfill, the oldest and deepest zone of Garbage Mountain.

These are the stories that everyone asks about, the weird, the dark and the unusual discoveries mixed in with the refuse. But that's not what's revelatory about any landfill's contents, Big Mike says. The truly thought-provoking part of the business as he sees it is the endless tide of ordinary, everyday stuff streaming into the place, items that are not really trash at all: the boxes of perfectly new plastic bags, still on the roll, tossed because the logo on them was outdated. Or the truckloads of food that turn up daily, a good deal of it spoiled but much of it perfectly edible, some of it still packaged and brand-new, yet discarded as if it had no use. Clothes of all types, some worn and torn but others seemingly pristine, are common. There are whole cans of paint (a forbidden, toxic item in landfills, though they arrive mixed in the household trash, difficult to detect), trashed because someone didn't like the custom color that, once mixed, could not be returned to the paint store. And there is the furniture—tons of it, much of it ratty and too far gone, but a surprising amount of it perfectly serviceable, at least until those chairs and couches and coffee tables meet Big Mike's BOMAG, the great democratizer of trash. Finally, poignantly, there are the old letters and photo albums, the vintage costume jewelry and ancient report cards in their brown manila envelopes. They are accompanied by sagging cardboard boxes labeled "important papers," though they aren't important to anyone anymore. This is the material that gets thrown away after someone dies. This is the world in which Big Mike plays king of the hill, and he sees what few people see, the fate of our most precious possessions, and how quickly, how easily, they become redefined as trash, deservedly or not. Someday, they might be treasures again, when the landfill reaches a great age and the broken but often still identifiable items within it become arti- facts awaiting an archaeologist's pick and brush. But that process takes a very long time, hundreds or thousands of years to turn mun- dane pieces of broken crockery into valuable artifacts. The more immediate process, the one Big Mike sees every day, is the one where treasures, our treasures of today, are used to construct a trash mountain.

Working there has changed him, he says, compelling him to think about how he and his family live, what they buy, what they waste. So many people buy so many things that they just throw away a year or two later—things that look great on a TV commercial, that promise to make life better or easier or more fun. Then those must-have products break or wear out, or simply wear out their welcome, and they enter Big Mike's domain.

The irony is that Big Mike's domain, with its unrivaled ability to hide seamlessly all that waste, empowers even more wasting. The landfill solution to garbage took away the slimy stench of the old throw-it-in-the-streets disposal, the smoking pall of the old incinerators, the noisome piggeries, the noxious reduction plants spewing out garbage grease, the ugly, seeping open dumps. It took away the obvious consequences of waste and eliminated the best incentives to be less wasteful. The rise of places like Puente Hills turned garbage from an ugly canker staring everyone in the face into a nearly invisible tumor, so easy to forget even as it swelled beneath the surface.

"It's such a waste," Big Mike observes. "More people should see what I see here, where everything that's advertised on TV ends up, sooner or later, and a lot sooner than most people think."

THE GOLDEN age of television and mass media marketing has been alternately celebrated and condemned for the last half a century for its unprecedented impact on society and culture. Yet one of its most enduring effects—helping bring about an American trash tsunami—is rarely put on the list of mass media goods and evils.

Not that the connection is disputed: Leaders of the industry during its earliest days admitted as much, describing their mission in life as persuading American men, women and children to throw away perfectly good things in order to buy replacements promoted as bigger, bolder and better. The senior editor of *Sales Management* magazine chronicled this when he wrote in 1960 that American companies and media outlets were working together to "create a brand new breed of super customers." A popular media journal of the day, *Printers Ink*, went further, suggesting the mission of marketers had to be centered on the fact that "wearing things out does not produce prosperity, but buying things does . . . Any plan that increases consumption is justifiable." Even President Eisenhower was caught up in the fervor, suggesting that shopping was tantamount to a patriotic act. When asked at a press conference what he thought people should buy to bolster the economy during a brief, mild recession, the president responded, "Buy anything."

And what a vehicle had emerged to persuade Americans to adopt this buy-more mission. For the first time ever, visually compelling moving, talking images were being beamed directly into their homes, free of charge, though not free of commercial messages—a transformative moment that rapidly shifted American popular culture into a round-the-clock tool for selling things. A new marketing industry began speaking about television viewers as a "captive audience" in whom it could instill "artificial" and "induced" needs—those are the terms they casually tossed about—for products no one had ever before considered a necessity. Their tirelessly upbeat portrait of American prosperity, the good life and "progress" made it quite clear that the American Dream was best achieved through buying the latest and greatest cars (preferably several per household), toothpaste and gadgets of all sorts.

This was the moment in which the Depression-era version of the American Dream—which held that hard work, diligent saving and conserving resources paved the road to the good life—began to fade, surpassed by the notion that the highest expression and measurement of the American Dream lay in material wealth itself, the acquisition of stuff. This was the moment when the perceived power to move nations and economies shifted from the ideal of the

American citizen to the reality of the American consumer. The phrase "vote with your wallet" entered the public consciousness in this era, elevating the act of spending money from unfortunate necessity to civic virtue.

This change did not happen in secret, unnoticed and unremarked upon. On the contrary, visionaries at the time saw clearly what was about to unfold and called it out—or, more often, celebrated it. The quote that best sums up the economic ethos of the day comes from J. Gordon Lippincott, an engineer by training, who pioneered the modern fields of product design and corporate branding. His marketing and design company's creations included the iconic Campbell's Soup label, Chrysler's "Pentastar" emblem, Betty Crocker's trademark spoon and the distinctive, stylized "G" of the General Mills cereal logo. Lippincott made this telling observation in 1947:

> Our willingness to part with something before it is completely worn out is a phenomenon noticeable in no other society in history . . . It is soundly based on our economy of abundance. It must be further nurtured even though it runs contrary to one of the oldest inbred laws of humanity, the law of thrift.

Lippincott's assertions would have sparked doubts about his sanity had they been uttered in Colonel Waring's day at the turn of the last century, when waste was public enemy number one, or in 1932, as the Great Depression crushed working Americans and the amount of trash generated by families reached record lows because every scrap of wood, paper and food was precious, and every tool and product had to be used, reused and repaired many times before anyone would consider consigning it to the trash. But Lippincott's era came in the very different boom times that followed victory in World War II, with America the last great power left standing and whole. His words became a marketer's battle cry signaling a break with the past. He plainly acknowledged that the disposable, thrift-less lifestyle he championed ran counter to the most basic human instincts and recent world history, yet he urged its embrace as part of the postwar belief that America—its resources, its wealth, its potential—had reached a state that had no real limits. What does waste matter in the America of Lippincott's imagination, with its infinite "economy of abundance"? The more we waste, in his view, the more stuff his clients could sell, the more consumers would buy, and the more prosperous America would become. *Failure* to waste was the enemy. If only Americans would waste more, they would boost production, enrich businesses and create more jobs—that was Lippincott's vision, and it was embraced as a marketing mantra by his colleagues, and then by most of the nation. The only way to defeat such an economy, Lippincott argued, was for consumers to lose their desire to consume in ever greater quantities.

His life's work, like that of the marketing and design industries he helped create and lead, was dedicated to preventing that from happening, to erase thrift as a quintessential American virtue, and replace it with conspicuous consumption powered by a kind of magical thinking, in which the well would never go dry, the bubble would never burst, oil and all forms of energy would grow cheaper and more plentiful with time, and the landfill would never fill up.

Lippincott is a fascinating historical figure, erudite, brilliant, seemingly easygoing. But there was nothing easygoing about his approach to reinventing the American consumer. With a team of 130 psychologists, sociologists, anthropologists and industrial designers backing him up, he mastered the art of making a product or a company or a concept appear to be something it was not. This was not the flimflam of the confidence man or the deceitful quackery of the snake oil salesman, but the art of the spinmeister—not hiding or breaking the truth so much as redefining it. He was helping to rewrite America's mythology. Eastern Airlines' noisy planes became "Whisperjets." U.S. Rubber, whose tires sold poorly abroad because of anti-American sentiment, became "Uniroyal." The gas station chain with an eye-glazing name and a reputation for poor service, Cities Service Oil Company, became the peppy "CitGo." Lippincott was so successful at designing to redefine that he was hired by the U.S. Navy to reimagine the interior of the Nautilus nuclear submarine to make its cramped quarters appear spacious, and by the U.S. Treasury Department to make the Internal Revenue Service appear more user-friendly (his one great failure, it seems, as the IRS apparently declined to change its name to something kinder and gentler). The rebranding, design and corporate identity industry Lippincott helped create succeeded beyond his wildest expectations. He helped make short-lived, disposable products and rising amounts of consumer waste appear to be virtues. He didn't hide the shortcomings previous generations would have identified in these products. He celebrated them.

It's no coincidence that the notion of saving up to buy something (and earning interest in the process) started losing ground in his era, supplanted by the then-new phenomenon of credit cards and borrowing to buy (and *paying* interest in the process). Political leaders who once fretted that Americans didn't save enough compared to the citizens of other wealthy nations, a disparity that still holds true today, eventually stopped complaining as the credit industry became a major force in the rise and fall of the American economy (and a major source of political campaign donations as well, more than $32 million a year leading up to the current financial crisis). The normalization of what was once unthinkable, even shameful— large amounts of consumer debt—added decades of longevity to the Lippincott vision of throwing out and buying more at an ever-increasing rate in order to manufacture boom times. Average household credit card debt topped the landmark of $10,000 in 2006, a hundredfold increase over the average consumer debt in the 1960s. One consequence: Much of the material buried in landfills in recent years was bought with those same credit cards, leading to the quintessentially American practice of consumers continuing to pay, sometimes for years, for purchases after they become trash.

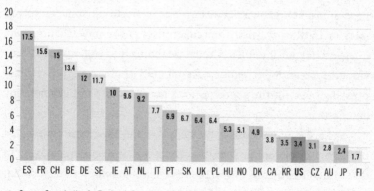

Household Savings as Percentage of Income, 2010

Spain (ES)
Belgium (BE)
Ireland (IE)
Italy (IT)
United Kingdom (UK)
Norway (NO)
Korea (KR)
Australia (AU)

France (FR)
Germany (DE)
Austria (AT)
Portugal (PT)
Poland (PL)
Denmark (DK)
USA (US)
Japan (JP)

Switzerland (CH)
Sweden (SE)
Netherlands (NL)
Slovak Republic (SK)
Hungary (HU)
Canada (CA)
Czech Republic (CZ)
Finland (FI)

Source: Organization for Economic Cooperation and Development

There were others at the time of this transformation who took an opposing view, who urged the country to reject "a society built on trash and waste," as the journalist and best-selling author Vance Packard saw it. Packard, whose first book had created a sensation in 1958 by citing examples of subliminal advertising and other manipulative imagery he accused the Lippincotts of the world of employing, wrote a prophetic follow-up in 1960 called *The Waste Makers*. In it, he accused his industry and marketing critics of sparking a crisis of excess and waste that would exhaust both nation and nature, until future Americans were forced by scarcity to "mine old forgotten garbage dumps" to recover squandered resources.

"Wastefulness has become a part of the American way of life," Packard wrote. "Some marketing experts have been announcing that the average citizen will have to step up his buying by nearly 50 per cent in the next dozen years, or the economy will sicken. In a mere decade, advertising men assert, United States citizens will have to improve their level of consumption as much as their forebears had managed to do in the two hundred years from Colonial times to 1939 . . . The people of the United States are in a sense becoming a nation on a tiger. They must learn to consume more and more or, they are warned, their magnificent economic machine may turn and devour them. They must be induced to step up their individual consumption higher and higher, whether they have any pressing need for the goods or not. Their ever-expanding economy demands it."

Packard couldn't foresee the details of this buy-more future, just the general momentum of the times sweeping the country in that direction. His was one of the first popular voices to suggest that growth, Lippincott's economy of abundance, had limits, and that

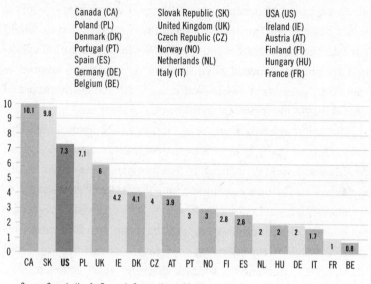

Household Credit Card Debt as Percentage of Assets, 2008

Canada (CA)	Slovak Republic (SK)	USA (US)
Poland (PL)	United Kingdom (UK)	Ireland (IE)
Denmark (DK)	Czech Republic (CZ)	Austria (AT)
Portugal (PT)	Norway (NO)	Finland (FI)
Spain (ES)	Netherlands (NL)	Hungary (HU)
Germany (DE)	Italy (IT)	France (FR)
Belgium (BE)		

Source: Organization for Economic Cooperation and Development

resources would eventually be used up. A prudent country, he argued a half century ago, should start planning to shift its economy away from consumption and rapidly paced planned obsolescence, and base it more on conservation, durability and elimination of waste before it was too late.

This view was mockingly dismissed at the time as Luddite pessimism. But not even Packard imagined Americans would achieve a 102-ton waste legacy. In his world, a household had one telephone, one television, one car, and the least of them were expected to last ten years or more—and that was a *prosperous* household. The idea that all members of a typical family, even young children, would someday have their own phones, and that Americans could pay several hundred dollars a month for this "necessity"—for cell phones that would become high-tech trash in two years or less—would have seemed so absurdly wasteful to Packard that he would have dismissed the idea as too fanciful. Yet this is the sort of evolution he predicted would have to occur as marketers labored to transform yesterday's waste and excess into today's normal and necessary. He just never thought it would go that far.

Packard's argument made sense and plenty of people bought his books. Not so many bought into his central premise, however, as Americans showed no inclination for embracing a retreat from consumerism. In 1960, his pessimistic, anti-materialistic views and prescriptions just could not compete with Lippincott's vision of endless abundance. People didn't want to see the waste—even Packard conceded as much. The overflowing trash cans were just more evidence of America's productivity. Indeed, fifty-one years later, in 2011, America's leaders were still looking at the world through Lippincott's eyes, even as they proved Packard correct by publicly stating that the best hope for pulling the country out of recession and unemployment—that is, the best way to get the tiger to stop biting us—would be for American consumers to shop the country out of trouble. Fifty years after Eisenhower, the message remained the same: Buy anything. Then throw it away and buy some more.

THIS RISE of consumerism and the new American Dream launched during television's golden age was accompanied by another trash-boosting trend—the plasticization of America.

Municipal waste was only .4 percent plastic by weight in 1960. Our trash cans had almost no plastic inside them back then, but that began to change rapidly. By the end of the sixties, plastic trash had increased sevenfold. By the year 2000, American households were throwing away sixty-three times as much plastic as in 1960, and nearly 11 percent of the stuff in our bins each week was made of synthetic "miracle" polymers. Plastic had become the second weightiest component of garbage next to paper, which, as daily newspapers in America have declined, is a shrinking category of waste.

These figures, which come from the Environmental Protection Agency, actually underestimate the impact of plastic ubiquity, because plastics are the lightest trash we make. If the components of American trash were measured by volume or pieces rather than weight, plastic would rate as an even larger portion of what we throw away. Consider the plastic shopping bag, which didn't exist in 1960. By the year 2000, Americans consumed 100 billion of them a year, at an estimated cost to retailers of $4 billion—costs passed on to consumers. Most of these bags land in landfills, though many go astray—plastic bag waste was the second most common item of trash found on beaches during 2009's International Coastal Cleanup Day (with cigarette butts taking first-place honors). Their relatively small weight masks plastic bags' enormous impact.

The new age of plastic went hand in hand with another trash-making trend of the era: the rising popularity of disposable products. These appeared on the scene in rapid succession: the invention of Styrofoam by Dow Chemical in 1944; the plastic-lined paper cup in 1950; the first TV dinner (turkey, mashed potatoes and peas on a one-serving, one-use plate and package) from Swanson Foods in 1953; the invention in 1957 of high-density polyethylene plastic, which would be used in the now-common gallon milk jugs that displaced reusable glass bottles; the 1958 introduction by Bic of the first disposable pen, followed by the 1960 marketing of the first disposable razors. There was a wave of such products, touching every part of the household and daily life, from plastic bread bags to non-refillable aerosol cans to the original paper Dixie cups. Dixies were initially despised for their wastefulness at the dawn of the twentieth century, but they were eventually accepted as tools for disease prevention in lieu of the traditional public glasses and ladles. (The product's original name had been the "Health Kup.") All of these new materials and products not only supplanted longer-lasting ones that could be reused many times and thereby remain outside of the waste stream, they also introduced a number of synthetic and potentially toxic waste materials into our refrigerators, medicine chests, cupboards, oceans, town dumps, natural habitats and our bodies.

One of the biggest shifts toward the modern throwaway economy came from the ultimate artificial necessity, a food product with no nutritional value and plenty of unhealthful effects, from obesity to dental decay: soda. In 1964, Coca-Cola decided it was finished with reusable glass soda bottles and introduced the "one-way" container, intended to be thrown out even though it was still glass and inherently reusable. Consumers embraced the convenience. The soda company was freed from the chains of local bottling plants (and local workforces) where bottle returns were cleaned and refilled. Three years later, the invention by DuPont scientist Nathaniel Wyeth of a plastic soda bottle that did not explode overnight in the refrigerator as its experimental predecessors had done—the polyethylene terephthalate (PET) plastic container—led to the marketing of the now ubiquitous two-liter disposable soda bottle. It and its smaller relatives have dominated the soft-drink industry ever since, ending a half-century age of reusable glass containers. Decades of recycling efforts followed that have yet to equal 1960 levels of soda sustainability.

At the time of this switch to plastic, the soda manufacturers cited the convenience, simplicity, lower weight and lower cost (to the manufacturers, at least) that made disposable plastic bottles the preferred and sensible choice over glass. It was progress. The market had "spoken." The notion that one-use containers made of fossil fuels—containers that never decompose—would inevitably impose a substantial cost on taxpayers, ratepayers, local sanitation departments and the environment was not considered in the industry's cost-benefit analysis favoring the plastic soda bottle. The reason for that omission was simple: The beverage companies did not have to pay those costs; such costs were, to use the corporate term, "external," which is a nice way of saying someone else has to foot the bill for a company's business plan. The new consumer economy, in effect, encouraged and subsidized the creation of new and greater volumes of trash—which, in the case of the soda industry, amounts to just under 10 billion cases of soft drinks a year in the U.S. alone. How much trash is that? Soft drinks are America's favorite beverage, quaffed daily by half the U.S. population over two years old (70 percent of males age two to nineteen drink sugary beverages daily, with the same age group of females right behind at 60 percent).[1] Annual consumption exceeded 50 gallons for every person in the nation beginning in 2002, more than twice the rate of soda drinking in 1960. (This is also twice the current per capita soda consumption in the number two soda-guzzling nation, Ireland; three times the soda drunk by Germans; five times the French; and ten times the Japanese.)

Fifty gallons a person is the equivalent of 160 billion 12-ounce soda cans, bottles and paper cups a year to be hauled to the curb.

NEW PRODUCTS and trends alone did not fuel the growing mounds of trash in postwar America. There was also the fading of old methods of garbage disposal. Another factor in this same era that boosted the trash flow, leaving sanitation officials in Los Angeles and many other communities scrambling for more dumps, was the demise of the garbage piggeries.

Until about 1960, piggeries still represented a major part of some cities' waste disposal strategies, not to mention the pork-production industry. Garbage piggeries may have been around nearly as long as the Republic itself, but they came into their own during World War I, when Washington war planners suggested that feeding garbage to pigs would conserve food. A similar boost occurred during World War II, when this was one of the forms of conservation practiced as part of the war effort, along with Victory Gardens, rubber roundups and scrap-metal drives.

Good data is hard to find on just how common it was for cities and towns to feed their garbage to swine. Perhaps the most authoritative source on the subject was Willard H. Wright, chief of zoology for the National Institutes of Health and a prolific author and public health advocate. He estimated in 1943 that there were at minimum 1.25 million garbage-fed pigs in the U.S. Based on U.S. Department of Agriculture figures, Wright believed that figure to be "very conservative," because it only counted hog farms that sent more than one hundred pigs to market every year. There were, according to Wright, a "very large number of persons who market less than 100 garbage-fed hogs per year."

Wright's own survey of municipal garbage disposal methods found that 53 percent of cities with populations of ten thousand or more fed part or all of their garbage to swine. Wright didn't survey smaller cities, but he asserted that the practice of feeding garbage to pigs was thought to be similarly pervasive in small-town America.

Los Angeles had a network of garbage feeding ranches to handle the edible portions of its waste in those days. By far the largest was Fontana Farms, fifty miles east of the city, which billed itself as the biggest hog farm in the world. Fontana at its height had a herd of sixty thousand pigs supplying a quarter of Southern California's ham and bacon. Feed consisted of daily shipments of 400 to 600 tons of Los Angeles garbage.

Though the war brought a resurgence in piggeries' popularity as a means of keeping heavy, rotting, vermin-attracting organic waste out of town dumps, that popularity did not extend to the use of garbage-fed pigs as human food. First there was the taste. Garbage-fed pigs, according to Wright, were poor in quality compared to conventionally fed porkers. "Garbage produces a soft oily pork," he wrote. "Cured pork and lard from such hogs are inferior in quality to that obtained from grain fed hogs . . . Most meat packers buy garbage fed hogs only at a discount." Wright also found that the "fattening value" of garbage had declined over the years, making it less attractive financially for farmers to adopt a trash diet for their swine: A ton of Los Angeles garbage in the 1930s could generate sixty-eight pounds of pork; ten years later, the slop put only thirty-one pounds on the hogs.

After the war, Wright advocated against the practice of feeding pigs raw garbage, gathering evidence that showed the meat from garbage-fed pigs was associated with higher human infection rates of the potentially deadly parasitic disease trichinosis. Garbage feed also was linked to several epidemics of vesicular exanthema, a fatal infection in swine similar to the bovine ailment known as hoof-and-mouth disease. As a result, by 1960, most states required the garbage fed to pigs to be cooked first to sterilize it. This proved so inconvenient and expensive that the use of piggeries as landfill alternatives quickly fell from favor. In the old days, Los Angeles was paid by pig farmers for its garbage. By the end, the city was paying the piggeries to take the slop, but it still wasn't enough, and they all but vanished nationwide by 1970.

Two trash-related inventions further contributed to this era's rising dependence on landfills: the compacting trash truck, and the green-plastic trash bag, introduced to consumers in 1960 by Union Carbide. In the past, open trash cans and uncompacted loads of garbage were easily picked over by scavengers, who made a living scoping out trash and pulling out recyclables and other reusable materials. Scavenging had long been considered an honorable vocation, one that provided economic opportunities for newly arrived immigrants in the U.S. in particular; the trash business in San Francisco, for example, was dominated in the early decades of the twentieth century by Italian immigrants who formed a network incorporated as the Scavengers Protective Association. The advent of dark green, opaque polyethylene trash bags, billed as a convenience, and garbage trucks that ground and crushed the trash into compacted masses for ease of disposal, had the unintended consequence of making scavenging far more difficult. No more lifting up the lid for a quick assessment of a trash can's contents—the bags hid everything. And once on the truck, everything was mixed and mashed, something that had not occurred with old-style wagons and open-bed trucks. Consequently, more material than ever ended up in the landfill instead of back in the manufacturing chain. Recyclables and food waste were hard to separate out. And the plastic bags themselves added to the waste stream as well.

The final factor driving the rise of more landfilling was the fall from favor of industrial-scale incinerators. Unlike the backyard burners, they had not been banned in many areas and there were quite a few still in use at that time, providing an alternative to land-fills. But two new federal laws hastened their closure: the Solid Waste Disposal Act of 1965, and the Clean Air Act of 1970. The old incinerators were too polluting to pass muster under the new regulations, while upgrading them to cleaner technologies was often prohibitively expensive. It seemed for a time that waste-to-energy plants would provide the next and best garbage disposal solutions, but it was not to be. Only the New England states—the U.S. region with the least amount of land available for new landfills—clung to incineration as a major solution, making a substantial push toward converting trash to fuel rather than fill. But even there, landfills remained dominant. The rest of the country turned to burial rather than cremation, and the age of the modern landfill began in full force.

FACING FAR more garbage than in the past, coupled with considerably fewer options for disposing of it, sanitation officials across the country began hunting for more landfill space to accommodate what was predicted to be a tidal wave of trash. In Los Angeles, a property in the Valley of the Dumps caught the attention of county sanitation engineers: a large and, at the time, well-known family dairy farm that had been operated for generations by the heirs of one of L.A.'s early land barons.

The Pellissier family's patriarch, Germain, had immigrated from France eighty years earlier, investing in sheep, cows and real estate. The land came cheap back then but would be worth a fortune for future generations of Pellissiers, particularly a rocky stretch on the edge of the city. This would eventually become the Miracle Mile portion of Wilshire Boulevard, in its mid-twentieth-century hey-day nicknamed "the Fifth Avenue of the West." The Pellissier name became part of the fabric of burgeoning Los Angeles, plastered on streets and buildings, the family money behind such landmarks as the historic art deco Wiltern Theater. Their dairy ranch remained an L.A. institution for decades (their Hazel the Cow, record-breaking milk producer, achieved minor local fame). The dairy farm finally shut down in the early seventies, as the value of the land had grown too great to consign to Holsteins in an era of seemingly insatiable demand for suburban houses and shopping centers—and dumps. In 1971, the county sanitation district bought a large portion of the Pellissiers' land along with the private dump that had been leasing space there. The county continued running the old dump as a small way station for garbage, while laying plans for an entirely new future for the place as the Puente Hills Landfill, the first big and thoroughly modern landfill in the region, with the new sanitation headquarters building built at its base.

Back then, three peaks dominated the property, part of the Puente Hills range on the southern border of the San Gabriel Valley. Canyons divided the hills and had been used for grazing the dairy cows. Residents in nearby Hacienda Heights could hear the cows coming home every day, a taste of country life in America's second-biggest cityscape. But the peaks were high enough to pose an aviation hazard for L.A.'s increasingly busy skies. In 1952, a North Continent Airlines twin-engine Curtiss flying from New York to Burbank was diverted because of fog, and while approaching its alternative landing site, Los Angeles International Airport, it dipped ten feet too low. Its landing gear struck the fog-shrouded peak. The plane crashed and exploded, killing all twenty-nine passengers and crew members aboard.

A year later, the first landing of invaders from Mars took place in those same hills—in the 1953 film adaptation of the H. G. Wells classic *The War of the Worlds*.

Cold War fears of a more earthbound invasion were soon added to the mix of bucolic pasture and suburban foothill: In 1956, the tallest of the three hills overlooking Los Angeles was chosen by the Defense Department as a strategic high point for a new Nike missile installation, America's first missile defense program. Batteries of radar-guided missiles designed to shoot down Soviet Union bombers flying as high as fifty thousand feet were positioned on the hilltop in the fifties and sixties. Similar outposts were set up across the country, continually upgraded to match faster and higher bombers. The Nikes were all decommissioned by the mid-seventies once the main threat of nuclear attack shifted from manned aircraft to much faster and more numerous automated nuclear missiles. Today the launch silos are covered over by a community college campus. The radar station at the high point has been repurposed as an aviation radar relay accompanied by a conglomeration of cell phone towers. Only a narrow wooden guard shack remains of the old Nike outpost to mark the spot, sporting a small memorial plaque.

But this former missile site no longer stands tall above a deep canyon. The Puente Hills landfill has filled in the spaces between the three hills, absorbing them into the larger footprint of a single mound of trash. The old high point that a missile launch complex once occupied, defending Los Angles from above, is just a small bump on the big, broad plateau of Garbage Mountain.

That was not the original plan, never the bold vision—the landfill was supposed to be a minor part of a much bigger effort to remake America's waste future while also weaning the country from foreign oil dependence.

This is, it turns out, a familiar theme in the history of trash—massive landfills as unintended consequence. Fresh Kills, which previously held the title of world's largest active landfill, was originally presented to the public in 1947 as a brief, temporary solution to New York's trash woes while new waste-to-energy plants were constructed. The plants did not come, and Fresh Kills continued operating until 2003. A decade later, the slow conversion of its polluted landscape into parkland is still under way. Philadelphia's leaders likewise believed they could eliminate the city's garbage with waste-to-energy plants, but a homeless, wandering barge of toxic ash from Philadelphia became an international pariah and scandal that killed that city's vision, too.

Puente Hills, destined to become America's biggest active landfill, has turned out to be as much cautionary tale as it is engineering achievement. It demonstrates once again how so many components of the modern waste-management system began as little more than a backup plan, an accident. The 102-ton legacy and the landfills that now constrain it are, bottom line, the unintended consequence of Lippincott's magical economy of abundance, superimposed on the American Dream by brilliant marketing designed to persuade people to accept the patently unsustainable as common sense.

4 THE LAST AND FUTURE KINGDOM

DAVID STEINER LOVES LANDFILLS. HE IS POSI-
tively poetic on the subject. "Landfills are amazing resources," he
says. "They're not just holes filled with trash. They're not the problem. They're part of the solution."

Okay, yes, he kind of has to say that. He is, after all, CEO of the
world's largest trash company, possessed of 22 million customers,
fifty thousand employees, 273 active landfills, $12 billion in reve-
nues and the most literal name in the garbage business: Waste
Management, Inc. He is America's king of trash. For him, trash is
money, and buried trash really is buried treasure, and that's the way

he wants it to stay. When a consultant he hired a few years ago to help chart a future course for the business suggested times were changing, and maybe the company needed to start redefining itself not as a master of waste but as Materials Management, Inc., Steiner dismissed the notion as rank heresy. "There will always be waste," he said confidently. "And we'll always need landfills. That's the core of our business."

A few short years later the idea still makes him emit a reflexive, nervous laugh. But he concedes that the consultant was on to something after all, now that trash is no longer a growth business, now that these pesky green cities all over the map keep talking about "zero waste." Some of his best customers, from the city of San Francisco to the mighty retailer Walmart, are clamoring to "recapture the value" of trash and "close the loop" on manufacturing and waste. These are code words for radically curtailing waste and, where that's not possible, holding on to trash as a valuable resource rather than burying it in Steiner's landfills. There is a growing sentiment that the future of waste just might mean a future freed from waste.

"Someday we might pay customers for their trash, rather than the other way around," Steiner allows, reflecting on an everybody-wins future in which trash companies pay a bit for garbage as raw material, then make a fortune turning it into the building blocks of the consumer economy. "We're not there yet, but it could happen. A few years ago, you'd never hear me say that."

If trash really *is* treasure, the king of trash's big challenge now is to figure out how to make this garbage-to-materials makeover work for his company, rather than bury it.

The irony behind this vision of sea change is that, a quarter century ago, it was Waste Management and its aggressive push to privatize the nation's trash heaps that helped diffuse earlier attempts to move away from landfills and toward more sustainable choices. Momentum that had been gathering in the seventies and eighties for alternatives to burying the 102-ton legacy all but died then, which meant that profitable and increasingly privatized landfills remained the default choice for America's trash strategy. Although Waste Management had no direct hand in determining the fate of such publicly operated locations as Puente Hills, the company's success at keeping landfills center stage made the Garbage Mountain of Los Angeles all but inevitable. A shortage of suitable landfills had driven interest in alternatives, including serious discussions dating back to the seventies, of government mandates for new designs of products and packaging to reduce waste. One 1972 EPA report, entitled "Mission 5000," took both government and the private sector to task for applying the world's best technology to creating "an abundance of consumer goods" while failing to close the loop and reclaim the materials in those goods for new products once the old ones wore out.

"We failed to apply either modern technology or modern management to the ultimate disposition of this abundance . . . In the last few years, Americans have begun to recognize the enormity of the problems posed by our reckless generation and careless disposal of solid wastes. Now, at last, we are beginning to grapple with the difficult, long-range problems."[1]

The report predicted not an end to landfills, but a greater emphasis on extracting value from materials before they ended up buried in garbage mountains as both an environmental and an economic boon. Four decades later, we're still talking about doing that, rather than getting it done. One reason: The glut of landfill space Waste Management helped create in the last decades of the twentieth century shifted market forces back toward landfilling. Landfilling was comparatively cheap and easy, while switching to the close-the-loop system the EPA recommended so long ago would be risky and difficult, and involve unknown conversion costs. That's why the transformation Steiner is now contemplating remains a distant possibility—although, to his credit, it is at least a subject of discussion again. For many years, it was out of the question.

Waste Management, Inc., was a very different company before Steiner took the throne. It went from a mom-and-pop operation in Chicago to the fastest-growing and hottest investment prospect in the country, only to have scandal morph it into the Enron of trash (before there was an Enron), and it almost went belly-up. Now reborn, it's trying to achieve something that would once have been unimaginable: striking a balance between its lucrative landfill business and a new mandate from Steiner to go green.

The biggest garbage company in the world started with a Dutch immigrant named Harm Huizenga, who came to Chicago in 1893 during the busy—and trashy—time of the Chicago World's Fair. (What is it with epic developments in the history of trash and world's fairs?) Huizenga started hauling trash for $1.25 a wagonload. He gradually built a successful family business, Ace Scavenger Service, serving Chicago and its suburbs. In 1956, the company had fifteen trucks and revenues of $750,000 a year. When the head of the company—Harm's son, Tom Huizenga—died, management duties passed to Tom's son-in-law, Dean Buntrock, who began aggressively expanding the business by buying up other small trash haulers. Trouble started almost immediately, as the company's tough tactics led to lawsuits and claims by the Wisconsin attorney general that Ace was threatening to destroy competitors' trucks and harm family members unless they made room for Ace's expansion.

The Milwaukee courts imposed an injunction against the company for eight years.

But Buntrock, who was only twenty-four when he took over the business, was unfazed. He kept snapping up more companies, and in 1968 he merged Ace with another aggressively expanding garbage outfit based in Pompano Beach, Florida, Southern Sanitation Service. This scrappy garbage company was run by his father-in-law's nephew, H. Wayne Huizenga, who would go on to make billions from building Blockbuster Video, Auto Nation and other business ventures. In 1971, the two men formed their companies into Waste Management, Inc., with $5.5 million in annual revenue. Buntrock was CEO and Huizenga was president. They took the company public and used the proceeds to buy ninety other garbage companies in nine months, rolling them into a rapidly growing WMI. The two budding trash magnates were banking (correctly) that new federal rules on garbage, landfills and the environment would create massive opportunities for large trash companies with expertise and resources that could take over landfills and collection for cash-strapped cities. WMI's rapid expansion continued into toxic and hazardous wastes, trash-to-energy plants, recycling centers and international sanitation contracts with Saudi Arabia, Buenos Aires, Hong Kong, Caracas, Brisbane, Australia, and several cities in Europe. By the mid-eighties (when Huizenga retired from the company), revenue reached $2 billion; by the mid-nineties, it topped $10 billion, and WMI had become the undisputed waste and landfill leader of the world, the biggest trash company in history.

But during that time of rapid expansion, Waste Management was also accused of illegal toxic dumping, cited for violations in seven states, and was peppered with accusations of influence peddling and destroying evidence. The company paid out over $20 million in fines for environmental violations, plus a $30 million settlement for hazardous waste dumping in three states, followed by a $91 million judgment for a fraudulent landfill purchase in Alabama. A 1991 Greenpeace report said of WMI: "To create an empire, the company has mixed business acumen and foresight with strong doses of deception, corruption, and monopolism." The San Diego district attorney issued its own report after WMI courted the city's business, finding that the trash company had a history of environmental problems, bribery and death threats. And in 1998, the company was nearly destroyed by one of the nation's biggest accounting scandals ever, when Securities and Exchange Commission regulators indicted four top company officers, including Buntrock, for massive fraud and insider trading. The four eventually settled the charges for $25 million, most of which was paid by WMI, and without admitting guilt.

The scandal devastated the company, its stock value dropped $25 billion, and a smaller company, USA Waste Services, swept in and bought the crippled trash giant, though it retained the better-known WMI name. This was neither a smooth nor a successful transition; the company went through five CEOs in four years. Steiner joined the company as general counsel in 2001 under a new regime that set out to clean up the chaos left behind by the founders and the buyout.

By the time Steiner took over as CEO in 2004, the company's reputation as a polluting, shady outlaw had been transformed, as the rampant environmental and safety violations of the old regime faded. And under Steiner's reign, WMI has begun to refashion itself as the new darling of the sustainability movement, with more than one hundred power plants converting landfill gas to electricity—enough to power 1.1 million homes. That's more output than the entire U.S. solar industry in 2011. WMI also has fielded one thousand clean garbage trucks that use landfill gas as fuel.

But those green technologies, beneficial as they are, don't dispense with landfills—they rely on them. So they fit nicely inside WMI and Steiner's wheelhouse. Things would get much dicier for Waste Management if the country made a serious shift away from managing waste and toward managing materials in some way that didn't involve building garbage mountains with it. The most valuable thing Waste Management's old leaders amassed wasn't the fleets of trucks or the city contracts to pick up trash from the curb. It was the real estate—those 273 landfills, which can accommodate nearly 5 billion tons of future trash before they're full. Waste Management currently is on track to bury 120 million tons of trash a year in those landfills, so there's plenty of room for the future, which investors love, because that trash real estate will generate $1.3 billion in revenue just from letting people dump their garbage there.

So if landfills are the heart of the company's business, how can Steiner shift that? How can the CEO of WMI say burying trash in a landfill is a big waste, that we ought to be doing something better with that material? How could he replace a business model as obsolete when that was his company's entire business model? His question is one being repeated throughout the waste industry, just as it has been for decades. A less wasteful future in part will require successful, profitable companies such as WMI to cannibalize their own businesses, something few CEOs are willing to contemplate, much less undertake.

So Steiner is hedging his bets, gambling on evolution more than revolution. He has shifted his investment strategy from acquiring ever more landfill space to also buying up small companies that are developing new technologies and techniques for extracting valuable chemicals, minerals and products from trash. The goal is to make recycling more effective and profitable. Just finding more efficient ways to sort and separate recyclables that currently get landfilled would be a huge breakthrough for both the environment and the company's bottom line. This may not sound all that radical, except this strategy, if it works, could make a majority of landfills obsolete over time. And remember, David Steiner really does love landfills. But he also loves the potential $10 billion in new revenues WMI could earn if it could capture the true value of the materials locked inside the trash his company collects and buries every year.

Steiner is one of the nice guys of the big-business world, a plain-talking country lawyer in his demeanor, who jokes that when he was (unexpectedly, he claims) handed the CEO job, he felt like going to the bookstore to see if someone had written *CEO-ing for Dummies*. His balancing act has turned into one of the trickiest high-wire performances in the business world, but the payoffs could be huge. Two of the companies now in the WMI stable are working on new processes that could turn trash into a gasoline substitute for cars and trucks far more efficiently and cheaply than existing methods of converting garbage to fuel. Assuming the new methods can be done on a large scale, it's possible that WMI could pull more than $200 worth of synthetic gasoline from a ton of trash.

Steiner believes that would be a game changer for waste management. Or, he jokes, would it then be materials management? Of course, even if the fuel conversion succeeds wildly, there's still not enough trash in the country to make the U.S. energy independent. "I see it as creating lemonade from lemons," Steiner says.

For now, the big money remains invested in landfills, if for no other reason than the lessons of garbage history, which suggest that betting on trash revolutions doesn't pay. Just ask the visionaries at Puente Hills, who are still smarting from their attempt three decades ago to reinvent Americans' relationship with their garbage.

IN 1983, the trash agency known as the Los Angeles Sanitation Districts applied for permits under the California Environmental Quality Act to expand its small trash heap in the Valley of the Dumps into a modern waste and energy facility. It was envisioned as a new type of garbage solution that would combine a network of high-tech power plants to convert the daily flow of garbage into electricity with a sorting and recycling center and a landfill to contain what was left over once the reclamation and generation was done. The landfill was supposed to be the least of it—a hill, not a mountain. Similar plans were being considered throughout the country at the time. Had they moved forward, they would have transformed the business of trash in America, making management of materials rather than trash burial the main purpose. Public landfills like Puente Hills and private enterprises such as Waste Management would have evolved in a completely different way. Our trash—and the path to our current 102-ton legacy—would have taken a different direction as well. But it was not to be.

This vision arose originally from a general trash and energy panic in the U.S. that began in the seventies and continued into the eighties. First there were the oil embargoes, gas lines and overall energy crisis of the era. These events had sparked a brief, federally backed renaissance for renewable energy sources (begun by Jimmy Carter, killed by Ronald Reagan) intended to ease the national and economic security nightmare of foreign fossil fuel dependence. Trash seemed like a ready, plentiful and affordable alternative fuel supply to bolster the effort.

Then there were mounting worries over the safety of the nation's aged, largely unregulated dumps—numbering about eight thousand active in the early eighties, with another twenty thousand closed landfills sitting and stewing. Many of the working dumps were rapidly reaching capacity and subject to toxic leaks, methane explosions, foul odors and contamination of drinking-water supplies as the chemical soup deep inside seeped into even deeper aquifers. Trash, sometimes with hazardous chemical waste mixed in, had been buried carelessly all over the country for decades without installing plastic barriers and other protections now deemed essential to containing landfill pollution. The result was a number of dire threats such as the Love Canal scandal in the 1970s, when a neighborhood near Niagara Falls, New York, was shown to have been built atop a toxic chemical disposal site from the 1950s, leading to a rash of birth defects, miscarriages, infections and other ailments. When Congress created the "Superfund" program in 1980 to clean up such toxic hazards nationwide, one hundred out of the first eight hundred most contaminated locations in the country were municipal landfills.

With such an array of fresh fears in the air, cities across America were having trouble finding suitable sites for new landfills. When they could find a good spot to bury garbage, community opposition got in the way—getting a new dump up and running was proving extremely tough. So why, the waste managers began to ask, should we endure so many problems burying garbage when much of it could be burned to make steam to drive generators that could power cities? We could kill, or at least blunt, two threats to our environment, security and prosperity with one elegant solution: waste-to-energy.

Those were different times marked by relative bipartisan environmental concern, an era in which the landmark Endangered Species Act passed unanimously in the U.S. Senate and the Environmental Protection Agency was signed into law by a conservative Republican president. The idea of burning massive amounts of garbage to generate a form of alternative (and somewhat greener) energy appealed to a broad cross-section of businessmen, politicians, activists and even some environmental groups. Industry leaders asserted that they could clean up the toxic and particulate emissions that bedeviled old-time, soot-spewing incinerators and still make it a profitable endeavor. Advocates asserted it would simplify the sorting necessary for recycling and so complement efforts to recycle more materials. (Others, however, feared it would do the opposite, since recyclable plastic is ideal, and therefore very tempting, fuel.) No one back then worried about the additional carbon emissions these plants would release. Global warming wasn't on the radar, with public and scientific angst focused instead on the prospect of the Cold War becoming hot. An exchange of atomic weapons, scientists warned at the time, could initiate a worldwide "nuclear winter," a dim, starved and frigid modern ice age triggered by immense amounts of sun-blocking dust propelled into the atmosphere by atomic explosions. In that reality, the threat of seeping landfills despoiling land and water was deemed a far greater threat than burning trash to keep it from the dump.

This waste-to-energy idea gripped the entire state of California then, where leaders imagined the state as a world leader in turning trash into electrical treasure, and exporting its expertise globally. It was embraced in a big way in the Los Angeles area by the two main trash-dealing government entities of Southern California: the city of Los Angeles, which has its own dedicated sanitation fleet and landfills, and Puente Hills's owner, the L.A. Sanitation Districts, a

unique quasi-public agency that serves the other seventy-eight cities in Los Angeles County, from Agoura Hills to Malibu to Whittier
(whose mayors compose the agency's governing board). Each juris-
diction, city and county, simultaneously laid plans for massive trash-
fueled power plants as the solution to aging, leaking landfills that,
in those boom times for Southern California, were rapidly filling.

The first step in making these plans a reality in Los Angeles was
the Sanitation Districts' permit to expand the old dump on the
Pellissier ranch into a modern sanitary landfill, a process that led
in 1983 to the first of many public hearings. These hearings were
not cordial. It proved to be a very difficult time to seek consensus
on plotting the future of the Valley of the Dumps. Robert Bennett
had just been murdered, his body missing but presumed hidden in
the trash of Puente Hills, raising questions about safety, security
and oversight. In Sacramento, dire warnings of a severe budget crisis gripping the state became a daily ritual, making large capital
projects such as power plants a dicey proposition at best. And in
Washington, the Reagan administration's top environmental officials, including the head of the Superfund, were under investigation
for cover-ups and for being in the pocket of toxic polluters and
crooked waste-site operators—hardly a climate conducive to trust-
ing the plans and promises of landfill operators in Reagan's native
state. Residents in the neighboring communities marched to the
podium at public hearings to express their alarm about the expansion of the Puente Hills landfill. They were outraged to have received official assurances that there were no hazardous materials
or leaks there, only to learn later that there had been both. Now
they wanted the dump shut down and Los Angeles's growing flow
of garbage redirected to some remote location out in the desert,
rather than have it dumped and piled in the midst of a growing

community of more than fifty thousand residents adjacent to Puente Hills who had been there long before the landfill. "We didn't move here for this," was the oft-repeated slogan.

But the trash bosses assured them that the future of garbage wasn't about landfills—it was all about drawing clean energy from waste. And in the process, the volume of the trash would be reduced 95 percent. They'd barely need landfills after that. The whole country was going in this direction, they confidently predicted. Burying trash for eternity was old school. Lighting your houses with it made so much more sense.

First, though, the Sanitation Districts' leaders said they needed a ten-year permit to expand the dump near the Nike missile site and turn it into a modern landfill while they planned the new trash-to-energy future, then got it up and running. There'd be more hearings once those plans took shape in a few years. In the meantime, an innovative power plant was already being built to make the old dump's noxious emissions into electricity instead of simply burning it in flaring stations—giant Bunsen burners powered by garbage gas—as was done in the past. When dubious homeowners and local activists pointed to the dense fine print of the new Puente Hills permit proposal—the parts that suggested the newly modernized landfill could, if necessary, operate for thirty years and absorb 100 million tons of garbage without breaking a sweat—this was dismissed as a mere contingency. Nobody wanted that future, what one homeowner prophetically envisioned as "seventy stories of ugly." That would be a crazy squandering of resources, it was agreed. "We have a lot of money in this," the Sanitation Districts' spokesman said of the waste-to-energy plans, and it was true—the plants would cost three-quarters of a billion dollars to build, to be financed primarily with municipal bonds. "We're committed."

So the road map for the Puente Hills landfill of today was approved. The landfill got a ten-year lease on life through 1993. The evidence suggests that the waste-to-energy goals announced at the time were sincere rather than a ploy or a smoke screen, that those in charge really did expect to make the landfill part of Puente Hills virtually obsolete. Combined with similar proposals up and down the state, those plans would indeed have made California the world leader in generating energy from trash, and provided a new model for the rest of the nation to follow. But then public opposition to such plants turned out to be even more vehement than the sentiment against landfill expansion, and it was accompanied by a new laissez-faire politics that dismissed concerns over fossil fuel dependence and energy security as if the crises and embargoes of the seventies had never occurred. The combination slowly strangled the waste-to-energy plans in California and most of the rest of the country, and contingency plans became the *only* plans. Garbage Mountain was born.

The energy plan had been nothing if not ambitious. Puente Hills was to have been the site of the largest waste-to-energy plant in the world, capable of swallowing up to 10,000 tons of trash a day. The smokestack, which proponents of the plant promised would emit no visible plume, would have reached up to 450 feet high in order to make sure emissions blew up and away from the neighborhoods below. It would rise as tall as three Statues of Liberty standing on one another's heads. That image alone was enough to alarm the locals. They imagined a towering spire despoiling the foothills and the low-slung suburban skyline. This was anathema to men and women who still recalled the sounds of dairy herds lowing and shuffling by every morning. They didn't know in the early 1980s that Garbage Mountain would eventually rise up higher than any smokestack would have, blotting out a much bigger piece of the skyline without a trash-burning energy plant there to suck up the waste and give them power in return.

The Sanitation Districts' plan also called for seven other smaller plants, with capacities ranging as high as 4,000 tons of trash burned daily, to be placed in strategic locations designed to reduce the number of miles trash had to be hauled around the Los Angeles Basin. If all were built, their combined capacity would be great enough to handle all the trash then going to L.A. landfills, with ample extra capacity to handle future trash growth. There would be little more than ash left to bury at any landfill once the plants came on line—the garbage "crisis" would be solved for decades. And power for up to a half million homes could be squeezed out of that trash at the same time.

The city of Los Angeles, meanwhile, separately proposed three large-scale plants of its own to burn almost all of the city's garbage. This proposal, dubbed Project LANCER (a somewhat tortured derivation from Los Angeles City Energy Recovery), though not as ambitious as the county's plans, was better publicized and drew most of the national attention and debate. It had been conceived in the 1970s as concerns mounted over the use of the city's remote landfills in the scenic Santa Monica Mountains. One plant alone was supposed to save 1.6 million miles of garbage truck travel to one of the city's distant garbage dumps. In comparison to the spewing diesel fumes from those trucks, the anticipated emissions from the state-of-the-art trash plant would, at least according to city officials, represent a net gain for the environment and the battle against smog.

Such cheery pronouncements were soon displaced in the headlines by the campaign to stop it all from happening. The specter of

Los Angeles's fight decades earlier over backyard and old-style industrial incinerators haunted the proceedings. The last time Los
Angeles fired up a trash incinerator had been in 1947, when the
thick, black smoke pouring out looked like the sickly plume of a
forest fire so big and so foul that it stopped traffic for miles. A city
councilman complained that his district was "inundated in ashes
like from a volcano," and the plant was soon shut down in favor
of . . . landfilling. Incineration was the past, sanitary landfills were
the future—that had been the line in the fifties and sixties. Now
people were bewildered: Hadn't city and county leaders spent years
convincing us that incinerators were the big evil, contributing to
smog, asthma, cancer and who knew what else? Now we're supposed to embrace all that again?

The backlash caught the sanitation engineers flat-footed. Four
hundred angry residents showed up at a 1985 city council meeting
in the nearby suburb of Duarte, where the leaders had initially ac-
quiesced to the waste-to-energy plants. Council members immediately reversed their position. Another 250 people showed up for the
next hearing in Puente Hills, condemning plans for the giant landfill incinerator envisioned there. Yes, they had been told waste-to-
energy was the future back in 1983. But they hadn't known then
that this future, this monster plant, the biggest in the world, would
be erected in *their* neighborhood. They had assumed it would be
built out in the desert somewhere, and Puente Hills would be phased
out or at least gradually shrunk until it closed when its permit
lapsed in 1993. Now they were being asked to accept a plant that
would be there for another twenty or thirty years, and their answer
was a resounding no. Protestors at the meeting bore signs saying,
"Dump the dump." They wore surgical masks to indicate their fears
about pollution should trash burning return in force to the Los An-

geles Basin. "Keep your ash out of the San Gabriel Valley," another protest sign proclaimed.

Similar complaints about plant location, about the city of Los Angeles trying to saddle the poorest parts of town with everyone's burning garbage, and concerns that the pollution solutions were not as effective as waste-to-energy proponents had claimed, all served to increase public opinion against trash burning for power everywhere it had been proposed for Los Angeles. Dioxins were again a huge concern, with opponents warning that the smokestacks would be spewing the same potent carcinogens that had made the military defoliant Agent Orange so harmful, something the sanitation officials vigorously disputed. It didn't help that Sweden, a leader in the waste-to-energy industry, had put a moratorium on building the plants at that time because of dioxin concerns. The moratorium was later lifted once new emissions controls were put in place, but the impression that the technology was dirty and dangerous lingered.

With hundreds of angry voters showing up at city council meetings and public hearings, it didn't take long for city council members in the small cities ringing Los Angeles to gauge the political winds and join the smokestack opponents rather than risk being voted out of office. Soon the opposition included sixteen cities served by the Sanitation Districts, along with two congressmen and a passel of state legislators. With that, trash power was dead. All but two tiny demonstration plants already under construction were canceled, reducing waste-to-energy to a hobby rather than a solution in Los Angeles. This mirrored developments around the state and most parts of the country.

And so Puente Hills became the go-to place for burying, rather than burning, garbage. The Sanitation Districts made do with collecting landfill methane to generate power, a process about half as efficient as burning the trash, producing far less electricity and doing nothing to reduce the volume of material going into the landfill.

At the time, the citizens who opposed the plans for Puente Hills had said they weren't necessarily against waste-to-energy "done right." They just thought it should be done in some remote location, far away from the city—away from their homes. Then the unsightly smokestack, the question of emissions, the flow of trucks filled with trash—all would be out of sight and out of mind, the way trash is supposed to be. "If we have to burn garbage, let's put it on a train and take it out to the desert," a leader of the anti-trash-burning coalition said. "It may cost us five or ten dollars more per person a month, but it's worth it. They shouldn't be built in a metropolitan area." This statement proved to be quite prescient in one way—the trash train would one day be chosen as L.A.'s trash solution—but it also reflected just how poorly the opponents of waste-to-energy understood the economics and logistics of trash.

The neighbors of Puente Hills thought they had scored a victory, but they had only traded a power plant for an even bigger trash mountain. And when 1993 rolled around, they were outraged anew when, instead of closing down as they expected, the landfill was allowed to expand and extend its life another two decades, to 2013. Sanitation officials couldn't resist a bit of schadenfreude at that, for the neighborhood opposition to the power plants had left the county with no other option but to ramp up trash burial at Puente Hills. There simply was no other place for the garbage to go at that point. The protesters had, in effect, made sure that the biggest landfill in the country would be in their backyard for decades. And so the tradition of creating waste-management systems through miscalculation continued.

Because of its convenient location, because the Sanitation Districts are a public agency with no need to amass profits, and because of savings from generating power on-site, Puente Hills became the most affordable place in California to dump trash. For many years it charged cities (the same cities that owned and governed the Sanitation Districts) as well as private trash collection companies and everyone else who needed to dispose of waste just $18 a ton to dump. This was half of what some other public and private facilities charged in Southern California. During boom times through the nineties and up until the recession, garbage trucks would line up at sunrise for the privilege of tipping their loads at Puente Hills, and by eleven in the morning, the gates would have to be shut, as the operating permit limits daily intake to 13,000 tons. Big Mike and his colleagues had to scramble to keep up with the constant flow of garbage at that pace.

Even charging below-market rates, Puente Hills took in more money than it could spend. By 2011, it had salted away a quarter billion dollars to pay for the next trash solution in Los Angeles. This is what made Puente Hills the envy of the landfill industry. There was even enough income to skim one dollar from every ton of earnings and designate it for preserving wildlands next to the landfill. No other active landfill in the country has nearly four thousand contiguous acres of hiking trails, parkland and wildlife preserve abutting a massive garbage mountain. The preserve is so huge and has attracted so much wildlife that the conservation authority created to run it has hired a full-time ecologist.

The dump and power plant opponents couldn't kill the landfill in 1993, or again in 2003, though they tried each time the ten-year permit came up for renewal. But they secured one other victory in addition to killing waste-to-energy: 2013, Puente Hills's thirtieth birthday, was designated as the irreversible drop-dead date for the landfill. The trash train plan advocated by the opponents is supposed to come on line then, and L.A. garbage is supposed to hit the rails.

But there is a complication. Those who wanted the landfill moved, and who imagined it would be only slightly more costly to railroad the garbage out of town, were wrong. It is a *lot* more expensive. Transferring trash from a new rail depot at Puente Hills to the Sanitation Districts' newly purchased former gold mine two hundred miles away in the desert of Imperial County will cost $80 a ton, more than four times what it costs to bury trash at Puente Hills. The Sanitation Districts can use its war chest to subsidize a lower price for the trash train, but even so, at a cut-rate price of $50 or $60 a ton, waste by rail will be more expensive than the power plants would have been, and far more expensive than Waste Management, Inc.'s private landfills in Los Angeles that, with the recession reducing trash flow, would be happy to take on the garbage now going to Puente Hills. It's unclear, in a tough economy, if this waste-by-rail plan can succeed, leaving Puente Hills and its neighbors in trash limbo. And Los Angeles has an unexpected sort of trash crisis on its hands: It is supposed to open a new, very pricey mega-landfill in the desert, with four times the capacity of Puente Hills and a lifetime of no less than one hundred years—and it just might have no trash it can afford to put in it.

The dilemma has raised, once again, the specter of garbage crisis, and the equally long-lived question America has yet to answer well: Isn't there something better we can do with, or about, our trash?

Decades of landfilling have answered, at least in part, the first of the big three questions that must be answered to begin to wipe away our 102-ton legacy: What is the nature of our waste? We may badly underestimate how much stuff we're burying, but we do have a good handle on what it's made of. And we also know what it's worth—some $50 billion in value chucked each year, lost to us now, but waiting to be recovered if only we could somehow make the transition from waste management to materials management that the king of trash, Dave Steiner, dreams of. Yet every time we have approached a new paradigm for waste, we have turned away from it, dating all the way back to Colonel Waring and his White Wings, who arguably were better at reclaiming and recycling materials than we are today. We are still in thrall to J. Gordon Lippincott's brilliant warping of human instinct from thrift to a disposable abundance. That marketing man's sleight of hand still commands us, having redefined the American Dream so thoroughly that it is hard to envision an America in which no need or desire could justify the construction of a mountain made of garbage.

BIG MIKE climbs down from his BOMAG, another day of bending garbage to his will behind him. Gone are those 13,000-ton days of garbage and gates that closed at noon. Now the landfill stays open till five, and gets nowhere near the daily limit.

Back in the landfill boom days, a third of the trash was commercial, a third was building construction debris, and a third was household waste. With the bursting of the housing bubble, construction waste in 2009 became a trickle instead of a flood. With the economy in recession and Los Angeles unemployment at 12 percent, people were buying less, so the commercial and household trash slowed down, too. All told, the daily flow dipped to around 5,000 tons a day at Puente Hills. And though the pace picked up again in 2012, it still lagged behind the old 13,000 daily tons. In theory, the place could stay open for years beyond 2013 without filling up. Toward the end of 2012, yet another battle started over whether or not to close Garbage Mountain as promised—again.

"The thing is, it *will* stay open, no matter what happens," Big Mike says, looking at his dozer fondly. "We may not take any more trash after 2013, but we're not going anywhere. *I'm* not going anywhere. The work here will go on a long, long time."

The day Puente Hills accepts its last load of trash, whenever that is, is the day it enters its "Terminal Phase." The first several years of the end phase will be spent placing a final, thick cap over the landfill, permanently sealing what by then will be close to 150 million tons of trash. Then there will be landscaping, road-building, park conversion—the transformation of Garbage Mountain into some other purpose, all of which will require big machines and Big Mike's skills driving them. After that, there will still be years of maintenance and monitoring and repairs and retrenching and re-installing gas lines and keeping the power generation going. There will be a full-time staff there for a long time. How long? Consider Puente Hills's predecessor, the Palos Verdes Landfill not far from the Pacific Ocean, part of which now lies tucked beneath a lovely urban oasis, the South Coast Botanical Gardens, billed as "the jewel of the peninsula." The landfill stopped taking trash in 1980. Thirty-one years later, the Sanitation Districts still had a staff there, still maintained the landfill, still made electricity with the gas that still seeped up from its depths, albeit at a reduced rate. There are thousands of "closed" landfills that are similarly still maintained spread across the country. As Big Mike says, landfills are forever.

"I'm not worried about my job," he says. "In this business, there's always work."

WASTE Q & A

1. If every country consumed and threw away at the rate Americans do, how many planets' worth of resources would be required to meet the demand?

2. America is home to 4 percent of the world's children. What percentage of the world's toys do Americans buy and throw away?

3. How many plastic water bottles do Americans throw away a second?

4. How much food do Americans throw in the trash every year?

5. How many people could be fed with 5 percent of that wasted food?

6. How many of your food dollars are spent on packaging?

7. How much waste does the entire U.S. economy create to make a year's worth of food, fuel, resource extraction and products for one American?

8. How much of that total waste figure is recycled?

9. How much energy is wasted on junk mail?

10. How much of your life is spent opening and throwing away junk mail?

11. How many barrels of oil are used to make a year's worth of disposable plastic beverage bottles for Americans?

12. How many liters of water are needed to make one liter of bottled water?

13. How much disposable plastic wrap is made each year in America?

14. How many nonrecyclable Styrofoam cups do Americans throw away in a year?

15. How much plastic trash ends up in the ocean?

WASTE Q & A ANSWER KEY

1. Five planets
2. 40 percent
3. 694
4. 96 billion pounds
5. 4 million for a year
6. $1 out of every $11
7. Just under 1 million pounds (waste water not included)
8. 2 percent
9. One day's worth could heat 250,000 homes.
10. Eight months
11. 17 million
12. Three liters
13. Enough to shrink-wrap Texas
14. 25 billion, or enough cups to circle the earth 436 times
15. The United Nations estimates there are about 46,000 pieces of plastic trash per square mile of ocean, and that 5.6 million tons of plastic trash are dumped, blown or washed into the seven seas.

5 DOWN TO THE SEA IN CHIPS

THE 151-FOOT-TALL SHIP *KAISEI* SLUICED THROUGH the summery waters of the North Pacific with that soft, graceful roll that always says "home" to Mary Crowley. The teacher turned sea captain turned ocean charter entrepreneur turned environmental activist learned to sail at age four, and has felt more comfortable on a ship's deck than a city street ever since. The view from the *Kaisei*'s deck, however, was not so comforting, as the day's catch laid out beneath the billowing sails offered her a particularly ugly sight.

The *Kaisei*—the word means "ocean planet" in Japanese, and doubles as the name of Crowley's environmental nonprofit, Project

Kaisei—had sailed more than a thousand miles off the California Coast, far north and east of Hawaii, on a research voyage through one of the loneliest, emptiest stretches of the Pacific Ocean. It's part of a vast circular gyre of currents several times the size of the continental U.S., with more than three miles of water stretching down beneath the *Kaisei*, a place that Crowley says defines the term "middle of nowhere." Yet it had taken hardly any time at all to haul on board a crusty collection of floating detergent bottles, milk jugs, water containers, chunks of Styrofoam, an old bucket, lawn chairs, twisted pieces of lost fishing gear and derelict nets, a well-used toothbrush and a plentiful array of jagged and cracked pieces of plastic, none from a source readily identifiable.

But finding these big pieces of ocean trash was not the main source of Crowley's mounting despair, though she has known these waters for nearly forty years and sailed here back when they were truly blank and pristine and breathtaking. She knows this sort of trash is a huge problem, entangling and killing more than one hundred thousand marine mammals and an even larger number of seabirds—no one knows for sure how many. But what really alarmed her this day wasn't the trash she could see. It was what she *couldn't* see that troubled her, after the bottles, cups and other bobbing trash had been hauled out, and the mirror of water and foam appeared deep blue and clear, flashing by beneath sun and pale sky as she stared down from the railing. The spray felt and smelled as it always did, cold and salty, fresh and ancient. But the net trailing beside the *Kaisei* told her a very different story, revealing in sickly detail what could not be seen with the naked eye from the *Kaisei*'s deck.

After just fifteen minutes of being dragged through seemingly clear water, the fine mesh of the net was clogged and coated. As they hauled it in, Crowley could barely believe what she was seeing, even though this was what she and her crew of volunteers, sailors and scientists had sailed a thousand miles to locate and study.

Tiny pieces of plastic, a trashy confetti too small to see from the ship deck, had swirled through the water and into the net. The bits of plastic were all colors and shapes, jagged and smooth, flakes and pellets, making a veritable plastic noodle soup. A hundred similar trawls across twelve hundred miles pulled up plastic every time. Each one increased the sick feeling in Crowley's stomach. The ocean, she realized, had turned to plastic chowder.

The worst part, though, the part that left her fearing for the future of pretty much everything, came during the night trawls from a sister ship on this expedition, the Scripps Institution of Oceanography's *New Horizon* research vessel. That's when the nets were set for lantern fish, those small, luminescent plankton eaters that come up from six hundred feet or even deeper waters to feed on the surface at night. These globally ubiquitous, finger-sized fish are a critical part of the food chain, with a host of variations and species that together represent an estimated 65 percent of the biomass in the ocean. Larger fish feed on the lantern fish, and bigger fish prey on them, as well as seabirds and marine mammals, and on up the food chain, right up to the fish that people eat, that civilization has harvested and relied on since there's been something called civilization, and before that as well. That protein, that nourishment, that vast marine ecosystem—all of it depends on many trillions of healthy little lantern fish feeding on even greater numbers of tiny zooplankton.

Two Scripps scientists on the expedition team collected and dissected those fish to see what, if any, impact all that trash confetti has on them, given that some of the plastic bits are roughly the same size and shape as plankton. The researchers found more than

9 percent—nearly one in ten—of the fish had plastic in their digestive tracts. The plastic was floating right there with the plankton, and down the hatch it went.

This is bad news, but it's unclear just how bad. It's one thing for a percentage of fish to die from ingesting inert plastic. The problem is, the ocean receives all sorts of toxic pollutants, heavy metals and hazardous chemicals—from storm runoff, illegal dumping, sewage, ships, oil rigs and many other sources. Brittle old plastic particles can act like sponges for these toxins, becoming floating pockets of concentrated nasties. How much of this occurs, how much is absorbed into lantern fish bodies, and how much moves up the food chain toward us is a big unknown at the moment. The research is just getting started.

But this much is clear: The Scripps researchers found that the fish responsible for maintaining a significant part of the global food supply were eating potentially toxic plastic at an alarming rate—24,000 tons a year in the North Pacific alone. That's what scares Mary Crowley about trashed oceans: a lot of little fish with a lot of plastic in their guts, headed our way.

"Welcome to Pacific Garbage Patch," Crowley says. "We knew it was bad, but really, it's worse than we thought. Worse than anyone thought."

Crowley and her nonprofit are not the first to discover that the oceans have become a dumping place for plastic, nor are they the first to suggest that this threat to the marine environment might be growing worse and more dire over time. Nor has her group done the only research or even the leading research on the subject. They are doing important and sometimes breakthrough work in a field that is just beginning to draw a new generation of marine scientists, but there are other larger, more established organizations with more money and publicity at their disposal than Project Kaisei, which was founded by Crowley and two friends, a surfer and a sailor.

But Project Kaisei is unique in one regard. While other groups concentrate on studying the problem and advocating measures to stop the flow of plastic into ocean waters, Crowley's ultimate goal is to combine garbage-patch research with the hunt for safe methods to extract the plastic waste from the seas. Many if not most experts say this is impossible—there's too much plastic spread over too great an area to even contemplate taking on that job. The cost and the logistics would be overwhelming, even if the technology existed to remove plastic confetti from the ocean without also scooping up all manner of living things in the process, doing more harm than good. And no such technology exists.

Telling the woman who piloted a sailboat on Lake Michigan at age four that something she cares about can't be done proved to be a good way to provoke a reaction. She is pursuing ideas for converting oil spill cleanup booms and skimmers to plastic duty, and has brought several inventors out with her on the *Kaisei* to test original prototypes for extracting plastic from ocean waters. She insists cleaning up the mess has to be part of the solution.

"That's what finally got me involved," Crowley says. "I kept hearing that cleaning it up was impossible, that all we could do is keep it from getting worse. I figured someone had to try. Why not me? Why not try to find a way to heal the thing I love most? I think we can find a way."

LANDFILLS REMAIN America's go-to solution for capturing trash, but gauging their contents and their alternatives is only the beginning, not the end point, of understanding the nature of waste. What about the trash that escapes? What happens to the debris that is hurled to the side of the road, that blows off trains, that spills from loading chutes, that is swept into rivers and storm drains by wind and rain? And what about the below-the-radar waste stream that consumers never see or hear about, though it occurs on their behalf? Pre-production pellets of plastic—tiny beads called nurdles— are shipped by the billions every year all over the country and the world, so that they can be melted down and reformulated into infinite varieties of plastic and products. Each piece of this embryonic plastic is the size of a lentil, sometimes smaller, twenty-two thousand pellets to the pound (which means annual production in the U.S. exceeds a quadrillion nurdles). Their tiny size makes them easy to ship, almost like a liquid, readily poured into railroad tanker cars, their most common mode of transport. But that convenient nub of polymer is also easy to spill and next to impossible to pick up. These mini-plastics are notorious for slipping through cracks, seams and gaps during loading, shipping and unloading. Wind and rain take over, along with swooping birds that mistake them for tasty grubs or edible crumbs. Ten percent of the plastic debris found during beach cleanups is made up of these nurdles, according to Greenpeace. A survey of the forty-two-mile coastline of Orange County, California (home to Disneyland, Laguna Beach, Rick Warren and Surf City, U.S.A.), found that its renowned beaches had an estimated 100 million nurdles mixed in with the sand, shells and rocks. The pellets are such a common find at sea that their nickname has become "mermaids' tears."

There are, in short, a multitude of ways for trash to escape and plastic to go missing. Sometimes, this happens by design. Researchers at University College in Dublin discovered that a single garment made of synthetic fabric can shed up to 1,900 tiny plastic fibers with each wash, and these tiny bits are flushed down the drain. Facial and body scrubs, increasingly popular substitutes for bar soap, are another deliberate source of plastic in the wild, though consumers may not know that those tiny abrasive particles that scrub and invigorate the skin are, for the most part, tiny grains of plastic. But whether by accident or design, there is only one ultimate end point for this wild plastic trash: the greatest feature, the biggest surface, the deepest chasm, the broadest desert and the largest burial ground on the planet. It ends up in the ocean.

Of course, the ocean isn't a single destination, but many—and depending on the region, the season and the weather, trash entering the marine environment can end up in the water equivalent of a vault, a grinder, a pond or a conveyor belt. Because of the multitude of currents that move, churn and mix ocean waters, as well as the effects of winds, tides and the rotation of the earth, plastic trash that finds its way to the sea often embarks on a complex course and a very long trip not easily charted or understood. Predicting the voyage of a single piece of trash is a far dicier proposition than plotting a journey to the moon or Mars. The latter can be calculated in seconds on the average laptop. The former, due to the vast variables and incomplete knowledge of what happens when sea meets trash, could choke a supercomputer.

First of all, not all plastic trash swept, dropped or dumped into the ocean gets very far. Some, perhaps as much as half, sinks fairly quickly, as a number of very common disposable plastics are heavier than seawater and therefore can't float. You won't, for instance, see in the Pacific Garbage Patch a lot of plastic cola bottles, except for the occasional ones that are intact with air inside for buoyancy. Once broken up, the polyethylene terephthalate (better known as PET), which includes most soda, water, sports drink and juice bottles, sinks to the bottom or gets washed ashore. This can pose a huge problem for coastal environments and beaches, but it's not a factor in the deep-ocean garbage patch. On the other hand, milk jugs and plastic bags—or, rather, pieces of them—along with plastic bottle tops are a common find in the garbage patch, as the high-density polyethylene used to make them is less dense than water.

Bottom line: About half the plastic that gets into the ocean floats, which means it tends to be better traveled than the most ardent frequent flyer. Some currents have been observed to deposit plastic trash from the United States on the shores of Japan, a journey that can take up to seven years. Another current peppers remote beaches in Alaska with huge piles of trash, some of which is later sucked back out to sea and put back on a circular course around the Arctic Ocean, only to end up back at the same Alaskan beach years later (though smaller quantities migrate across the polar ice pack and find their way to the Atlantic). This phenomenon was first tracked by oceanographers who followed the courses of twenty-eight thousand plastic bath toys (ducks, beavers, frogs and turtles, bearing the "Friendly Floaties" brand) that were washed overboard from a cargo vessel during a storm in 1992. The colorful toys, eventually bleached white by the sun and sea, kept turning up for years on the same beaches, recovered by a worldwide network of volunteer beach-combers.

But there are other spiraling currents that seem to trap the plastic that enters their grasp without letting many pieces escape. These currents are called the five gyres, and they are located in the Indian Ocean, the South Atlantic, the North Atlantic, the South Pacific and the North Pacific. These are vast, constantly shifting areas of deep water that together encompass about 40 percent of the global ocean surface—which means the gyres cover more of the earth than all the dry land put together. Much of their volume is composed of the marine equivalents of desert—huge, empty, their diversity well hidden beneath the surface, although there are a scattering of zones within the gyres with incredibly rich sea life, as in the area around the Hawaiian Archipelago.

The mechanics of the gyres are complex. Multiple currents enter the gyres moving in opposite directions. The Coriolis effect—the force exerted by the rotation of the earth that causes water to circle a drain clockwise above the equator and counterclockwise below—induces the currents to converge and form a spiraling gyre. These zones also tend to have steady mild winds that augment, rather than disturb, the convergence. This slow spiral carries floating debris toward the center and tends to keep it there (given that there is no drain in the ocean for it to be sucked down through). Additional, ever-shifting concentrations of plastic and other debris have been observed on the edges of some of the gyres, too, particularly in the Pacific, where atmospheric high-pressure areas create persistent, ultra-calm conditions.

Researchers have found all five of the major gyres have higher concentrations of plastic than other parts of the ocean (although plastic confetti has been observed in waters and beaches all over the world). The North Pacific Gyre, home to the North Pacific Garbage Patch, occupies the zone of the Subtropical High between Hawaii and California. It is the largest and best studied of the gyres, though still fraught with unknowns. It is thought to be the trashiest, though this question is still being studied. Covering more than 20 million square miles, it is the largest ecosystem on earth—and therefore the planet's largest garbage dump.

In 1997, a sailor and ocean researcher based in Long Beach, California, was heading home from a boating race in Hawaii aboard

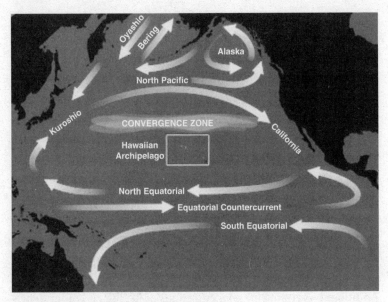

Source: National Oceanic and Atmospheric Administration

his vessel, the *Alguita*. Charles Moore's strategy during the race had been to avoid the waters of the gyre, as most sailors and fishermen do, because they are notorious for their low winds. Parts of this area of the Pacific have been called "the doldrums" for centuries because of their tendency to leave sailing vessels sitting still for days and weeks at a time. It's an area sailors have learned to skirt, a vast oceanic desert.

But the *Alguita* had powerful twin engines and plenty of fuel to supplement its sails and compensate for the gyre's lack of wind, and so Moore chose to take a shortcut through the doldrums. As the ship approached the gyre's center, he noticed the trash. First a little, then more. It was a transformative experience that Moore later wrote about in the journal *Natural History*:

I often struggle to find words that will communicate the vastness of the Pacific Ocean to people who have never been to sea. Day after day, *Alguita* was the only vehicle on a highway without landmarks, stretching from horizon to horizon. Yet as I gazed from the deck at the surface of what ought to have been a pristine ocean, I was confronted, as far as the eye could see, with the sight of plastic.

It seemed unbelievable, but I never found a clear spot. In the week it took to cross the subtropical high, no matter what time of day I looked, plastic debris was floating everywhere: bottles, bottle caps, wrappers, fragments.

After returning home, he contacted Seattle-based oceanographer Curtis Ebbesmeyer, who had become well known among ocean researchers for mapping currents by tracking the fate of the spilled Friendly Floaties bath toys. When Ebbesmeyer heard what Moore had seen, he said it made perfect sense that those zones would aggregate floating trash, and he dubbed it the Great Pacific Garbage Patch. The name was catchy and it stuck, although Moore came to regret it, as it misleadingly suggests the existence of some large and clearly visible trash island swirling around the Pacific rather than what it is: a soup bowl—or, as Moore once suggested, a "swirling sewer"—of barely visible particles circling endlessly.

Moore was the first person to bring the garbage patch into the public eye—and the first to devote his time and resources to researching it, with the *Alguita* outfitted as a research vessel and his family fortune, made in the oil business, fueling a foundation dedicated to tackling the problem of plastic pollution in the ocean. His first scientific paper, published in 1999, made what was then a shocking revelation to scientists who thought they understood ocean pollution: that in the Pacific gyre area, plastic fragments appeared to outweigh zooplankton by a factor of six to one. He followed up with a study that showed the waters off the coast of California had two and a half times more plastic than plankton, which was perhaps even scarier than the earlier study because those waters were not inside a gyre with currents that might be concentrating plastic debris.

In other words, Moore told the world, the plastic chowder is everywhere.

"No matter where you are, there's no getting over it, no getting away from it," he has said. "It's a plastic ocean now . . . We're putting everything in the ocean on a plastic diet."

Moore is passionate about continuing to publicize and study the garbage patch, and equally adamant about the solution: changing the way we live, removing disposable plastic from our daily lives. Shut off the supply, he says, and maybe the ocean can begin to heal. But trying to clean up what's already there? He reserves words such as "impossible" and "bullshit" for that. He's nothing if not candid.

And there's Mary Crowley's inspiration to found Project Kaisei: Moore's passion, his revelations. And his conviction that cleanup is bullshit. She begs to differ. Moore is half right about his solutions to the problem, she says. "We need to do both."

MARY CROWLEY is a trim, soft-spoken sixty-one-year-old sailor, her face weathered from years of salt and sea, her gait marred by a slight limp—not from a mishap in her element at sea, she's quick to say, but from a Christmas shopping expedition. A dark street, an awkward encounter with a curb and arms laden with presents produced a tumble that left her with multiple fractures to ankle and knee. The prognosis from her surgeon: In all likelihood, she wouldn't walk without a cane again. Capable of only one response to such a pronouncement, Crowley immediately set out to prove her doctors wrong. Months of painful therapy and exercise later, there was no cane for Crowley, no slowing down, no backing down. It's no different with plastic.

Mary Crowley is a classic "second act" baby boomer. After thirty-two years building a successful career and business, she has seized upon a new passion and pursuit with the same doggedness she applied to physical therapy. Championing the marine side of the ocean-versus-trash dilemma is her second act in life, her opportunity to give something back in the battle against her own 102-ton legacy.

When we spoke, she had just learned of an incident that horrified yet energized her, involving a scientist at a marine mammal center in the San Francisco Bay Area and a young sperm whale, only two years old, that had foundered and died. The scientist performed a necropsy to determine the cause of death, because the whale seemed outwardly healthy. She found 450 pounds of debris, most of it plastic, in the whale's digestive tract, just taking up space. The whale, its stomach full, had starved to death.

"This is a dreadful story," Crowley says. "Unfortunately, it's not an unusual story. If we're lucky, someday it might be."

It's not hard to trace the path that led her to take on this issue and the work of Project Kaisei. Some of her earliest memories of growing up in Chicago are of sailing with her family on Lake Michigan, her father, a state court judge, teaching her and her brother the ins and outs of piloting their small sloop. By age twelve she had devoured a series of books about sailing around the world and had started picturing herself at the helm; by high school graduation, the class yearbook was predicting her most likely career: sea captain.

After college and a major in psychology, she moved to Sausalito (where she is still based) and took jobs delivering yachts, sailing the boats to ports all over the world. By twenty-two, she had sailed more miles and visited more ports than she could count. That was when a crew member on a delivery to Tahiti told her about a job he had just accepted teaching aboard a tall ship out of Norway, a floating college program. She, too, ended up signing on to the floating faculty, teaching philosophy, psychology and navigation for a school year while earning her Norwegian seaman's papers. After that, she worked for five years for the then-new Oceanic Society, a marine environmental protection and exploration group. She ended up directing its expeditions programs, introducing hundreds of people to the sea, to sailing, to conservation and to far-flung destinations— and building a global network of maritime contacts and resources.

In 1979, all of her experiences at sea, in leading expeditions and captaining vessels large and small, culminated in her going into business for herself when she launched Ocean Voyages, her yacht charter and vacation company. Along the way, she married and divorced and had a daughter, Colleen, who sailed all over the world with her while growing up and is now a molecular biologist. And then, she recalls, after years at sea, she began to see the changes. She began to see the tide of trash washing in.

At first she would just talk about it with her fellow captains, grumbling about the pristine places they had long loved to visit that were no longer so pristine. The change was gradual at first, but undeniable, first a little trash here, a little there, then more, then a wave of it. The oceans she sailed thirty years ago, twenty years ago, even ten years ago had not been so trashed as they appeared now, she says. She knew she wanted to do something, but her course of action didn't crystallize until she grabbed on to the idea of researching the garbage patch with an eye toward finding ways to clean it up. An amorphous and huge problem suddenly became, to her at least, a clearly defined mission. Our health is tied to the health of the ocean—that, she knew, was axiomatic. "This," she says, "is a matter of survival."

Crowley partnered with two like-minded friends in founding Project Kaisei—Doug Woodring, an economist based in Hong Kong and expert swimmer and paddler, and George Orbelian, a real estate broker and surfboard designer. Different networks and skills, same love of the ocean: Together they brought in $600,000 in donations and grants to finance the first *Kaisei* expedition, the partnership with Scripps and its research vessel *New Horizon*. With the project's own science team aboard the *Kaisei*, the two ships together were able to cover twice as much of the gyre as would have been possible for Scripps alone, gathering water samples from more than 3,500 miles over three weeks. The result was a vast amount of data—still being analyzed—with which to begin to map the size and plastic concentrations of the Pacific Garbage Patch, and to measure the ingestion and toxicity of plastic eaten by nocturnal lantern fish and other sea creatures.

Then there was the science of cleanup to be tested, because Crowley wanted the trip to be about solving the problem as much as it was about studying it. This job fell to *Kaisei*'s consulting engineer, inventor Norton Smith, who arrived at the San Diego docks from Jacksonville, Oregon, with four different disassembled prototype plastic capture devices to haul aboard the tall ship. Mimicking nature in his designs, Smith named prototypes The Lagoon, The Beach, Sweep and Pyramid, after the objects or natural features they resembled.

Smith's design criteria were simple. First, there had to be a complete rethinking of how to capture the plastic. The standard manta tow nets that the two research vessels used to gather water samples and measure plastic concentrations were great scientific tools, but they would be terrible for large-scale cleanup operations because they capture great quantities of sea life. They are indiscriminate. Using that sort of capture would end up wreaking far more havoc than the plastic, killing the very organisms that were supposed to be saved from the dangers of plastic. Second, such a cleanup method would be fabulously expensive, requiring an enormous fleet of ships towing many thousands of nets for years, expending huge amounts of fuel. It would basically require the same energy and resources currently used for fishing, without any of the financial return. It has been suggested that the plastic gathered in this way could be used and even sold as fuel, but Smith's analysis found that, even assuming a completely efficient operation (an unrealistic assumption at best), only one-hundredth of the fuel needed could be generated from burning the ocean plastic. It would be a losing proposition of unprecedented magnitude, which is why so many smart, concerned people say cleanup is impossible.

The only way around these twin obstacles, Smith concluded, would be through inexpensive, passive plastic gathering devices using common materials and requiring very little expertise to assemble. And once deployed, they would have to be able to do their job without a fuel supply and without being towed by ships, but merely tended once or twice a day while the devices went about their work.

Out in the garbage patch, plastic bits swirling about them, Smith, Crowley and Smith's niece, Melanie, dove in the water to hook up what turned out to be the most successful of his prototypes, The Beach. In its design, Smith tried to re-create the physics of a typical inclined beach, in which plastic debris is readily washed ashore, but relatively few healthy sea creatures are beached. His device consists of a plywood inclined plane with a leading edge one foot below the water's surface and a trailing top edge that's about three inches above the water—basically a five-foot-wide floating boat ramp with plywood walls on either side. The Beach is anchored to a weighted parachute submerged twenty to forty feet below the surface, where currents in the gyre are stronger and tow the device through the water. At the surface, water breaks over the top end of The Beach, where a net is attached to capture the flow.

The *Kaisei* crew released the device and let it do its thing for nearly eleven hours, during which time the currents and parachute moved it more than three nautical miles. When The Beach was retrieved, Smith found that the net was full of small plastic particles and almost no sea life—the contraption worked.

An armada of such devices would be substantially less costly than more traditional, fuel-intensive methods of gathering the plastic, but would still be extremely time-consuming and expensive. The Pacific gyre would take about sixteen years and nearly half a billion dollars to clean up in this way, he figures—a daunting prospect at best. And he also makes clear, as does Crowley, that such a massive effort would quickly become a pointless exercise without something else even bigger happening at the same time: a worldwide reduction in disposable plastic garbage, and an end to the constant flow of plastic that goes missing every year, and ends up as marine chowder.

Crowley was still thrilled with the testing. For a first attempt, Smith's ideas showed promise, though she understands these sorts of technological solutions are a long way from being ready for prime time—if they are ever ready.

In the meantime, finding ways to encourage the removal of some of the largest and most dangerous pieces of ocean garbage—the tens of thousands (perhaps hundreds of thousands) of "ghost nets" adrift in the ocean—should be a top priority, Crowley says. These nets aggregate thousands of pounds of trash, they become caught on coral reefs, then break them apart during storms, and each one can entrap dozens, sometimes hundreds, of fish, birds and sea mammals. The *Kaisei* encountered quite a few on its voyage. Capturing derelict nets sometimes required most of the twenty-person crew to engage in a kind of tug-of-war as they hauled in the enormous, twisted lattices of plastic festooned with trash and barnacles on board, cutting parts of it with a torch in order to wrestle the thing out of the water, so weighted down was it with trash and dead creatures it had snared. There are thousands of tons of such abandoned ghost nets rolling through the gyre, giant marine death traps. No accurate global inventory exists for ghost nets, although the numbers even in small areas are staggering: three thousand estimated to be loose in Puget Sound alone; 1,800 removed from waters off Hawaii in recent years, without putting a dent in the problem.

"The Beach," *courtesy of Norton Smith, Project Kaisei*

Several pilot programs around the world have had success in paying fishermen to catch these nets in lieu of trawling depleted fisheries; Crowley advocates permanent and larger-scale programs to put fishermen to work undoing the damage ghost nets cause.

Beach cleanups also help, she says, because removing the trash from the surf cuts off a major source of "food" for the garbage patches.

This dovetails with the final part of Project Kaisei's mission: education and media, getting the word out about the problems of oceanic trash and plastic pollution. Crowley brought a documentary film crew, live bloggers and journalists with her aboard the *Kaisei* during the expedition, documenting the trash and the nascent efforts to combat it. Awareness, she says, is the best weapon against the trash, and the best goad toward action.

"I want everyone I can possibly reach to understand what we experienced on this voyage, what a very disturbing experience it was to be in the middle of the ocean, where one should be finding pristine oceanic wilderness, where there's nothing but ocean on all

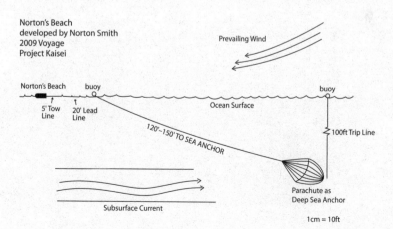

Norton's Beach
developed by Norton Smith
2009 Voyage
Project Kaisei

Prevailing Wind

Norton's Beach   buoy   buoy

5' Tow   20' Lead   Ocean Surface
Line   Line

120'~150' TO SEA ANCHOR   100ft Trip Line

Subsurface Current

Parachute as
Deep Sea Anchor

1cm = 10ft

your horizons, a place that to me is full of wonder, and you are see-ing our own garbage. You see laundry detergent bottles and bleach bottles, children's toys, toothbrushes, plastic buckets, storage con-tainers, packing straps. All this stuff out there in the middle of the ocean, it just makes me sick. And I want everyone to feel that, too."

It turned out this part wasn't so hard. She has found that ocean trash is a unique environmental issue. It is that rare green cause that transcends politics and ideology—once people see and under-stand it. Garbage floating on the waves, it seems, has the power to unite. Ten thousand visitors showed up at the docks in the days after the *Kaisei*'s return, eager to tour the ship and see the array of trash and ghost nets that the crew put on display on the deck, to learn about the gyre and to hear how that distant place was full of all of our trash.

"I've never talked to anyone who has seen the pictures or the video we've brought back, or who came to the ship to learn what it's really like out there, who then says, 'I don't care.' That's why I'm hopeful."

NERDS VS. NURDLES

THE SCIENTISTS TRYING TO FIGURE OUT THE jabberwock-sized problem of the gyre garbage patches tend to be characters. Miriam Goldstein is no exception.

Goldstein came to the work at Scripps after a post-college break from academia that included stints as a construction worker, an environmental consultant, a naturalist at New Hampshire's Mount Washington and a salesperson at a biological curiosity shop in Soho called Evolution. Now she represents a new generation of ocean researchers eager to launch their scientific careers by uncovering the extent and consequences of marine plastic pollution.

Goldstein is, she says, part of a new army of nerds taking on the legions of nurdles. "There's a lot we don't know yet, and it will take years of study to really get a handle on the extent of the problem and its impact," she says. "But we don't need to know everything to know that we should stop putting trash in the ocean. We already know that should stop."

She tends to see the state of the sea as the ultimate in societal heedlessness—an unintended and untended lab experiment run wild, in which the world finds out just what happens when we dump fifty years' worth of plastic into the ocean. Now, Goldstein says, it's time to assess the damage and figure out where to go from here. As part of that effort, she has been on extended sea voyages four times in less than two years, gathering data for Project Kaisei, Scripps, NOAA (National Oceanic and Atmospheric Administration, the most fitting acronym in government) and her own dissertation on the impact of plastic micro-debris in the North Pacific. Her work is part of the ground-floor research finally being done systematically on ocean trash after a decade of being left to a few capable but extremely shorthanded mavericks and gadflies.

In an ocean culture dominated by old salts with tan and craggy faces, the fair and freckled Goldstein seems the unlikeliest of oceanic heroes, quite the opposite of veteran sailor Mary Crowley, who thinks storm-swollen seas are a fun sort of challenge. The young scientist, by contrast, describes herself at sea as an "accomplished barfer." To her credit (or, according to her family, as evidence of her insanity), she charted her professional course despite knowing her landlubber tendencies, revealed during her very first sea voyage. This was a half-day whale-watching excursion off the coast of her native New Hampshire, which she organized at age ten, dragging her reluctant sister, brother and parents along as she pursued her passion.

"Let me just say, we're not really a very outdoorsy family," she recalls with a laugh. "The entire family spent the entire time barfing over the rail. We have pictures of the four of us lined up."

For her doctoral studies at the University of California, San Diego's Scripps Institution, she prudently planned to focus on coastal pollution, which would have safely limited her sea excursions to knee-deep wading into tide pools. That changed in 2008, however, when she started reading about the gyres, the garbage patch and ocean plastic—and how little we really knew about it. She proposed that the University of California Ship Fund devote some of its grant money and research vessel time to studying the matter, which the institute honchos soon agreed was a good idea, leading to the August 2009 SEAPLEX voyage (Scripps Environmental Accumulation of Plastic Expedition) to the garbage patch. The only caveat: Goldstein, queen of the tide pool explorers, had to take charge as chief scientist and see her idea through, an intimidating but unrefusable opportunity that knocked the twenty-five-year-old ocean scientist out of her wading boots for good. Initially planned for two weeks, the voyage was expanded to three when Project Kaisei offered to provide a second ship and more financing for an in-depth look at deep-sea ocean plastic.

The journey shattered Goldstein's expectations and ended up shocking her whole team despite weeks of burrowing through piles of reports, papers and news clippings she thought would prepare her for what lay ahead. She and her colleagues had spent long hours planning the trip, obsessing on how they were going to find the trash, and what they would do if they started roaming around this vast ocean desert without finding anything. It's a big ocean out there, they kept reminding one another. You can go out looking for something and the weeks can fly by—and then you come up empty. That was Goldstein's fear as the 170-foot *New Horizon* left port in San Diego. She knew marine biologists who went to sea time after time looking for certain organisms or feeding patterns or weather phenomena, and they just never found them. This was the oceanic equivalent of one of those space launches in which the parachute doesn't open or the radio goes dead or the probe drifts off into space without ever establishing contact with mission control. Things are spread out at sea, and the ocean seems to delight in frustrating scientists and crushing their attempts to uncover its mysteries.

So Goldstein and her colleagues were taken by surprise when it turned out to be all too easy to find the garbage at sea. As it happened, they simply set course for the gyre and the trash found them. They had been conditioned by press reports and the very name—Pacific Garbage *Patch*—to expect an actual patch, a visible aggregation of garbage, which news story after news story described as a kind of floating island of debris twice the size of Texas. But they did not find a bunch of trash in one place. What they found were high concentrations of small plastic bits spread across the entire 1,200 miles of ocean they traversed and trawled, finding plastic in every net. Jellyfish and sea slugs would come up in the net, swimming amid the plastic. Inside the jellies, plastic could be seen through their transparent bodies. It was far worse than they had imagined, not an island, but that damn plastic chowder. And it was everywhere.

"After days of endless plastic," Goldstein recalls, "we were all getting really depressed." But the prevalence of plastic had a silver lining. Finding it meant they had a good shot at understanding it. Three grueling, thrilling weeks followed of water sampling, manta tows (a special net shaped roughly like a manta ray deployed on the side of the ship), microscope work, plankton preserving (rotting plankton, Goldstein says, is not a smell you want to experience if at all possible) and captures of plastic debris small and large.

Goldstein's primary interest is how the unusual critters that live in the gyre, most of them small and many of them microscopic, interact with the debris and plastic in their midst. Do they peacefully coexist? Is it poisoning them? Do these added surfaces to cling to and lay eggs on—in an area of the sea where there is no land for a thousand miles—give a leg up to some creatures in the ecosystem at the expense of others? And what happens if those tiny crabs, barnacles and other opportunistic hitchhikers cling to a hunk of plastic and get swept by the gyre to a place where they don't belong? Nature's fragile balance, its chains of prey, predator and symbiont, could be altered by the plastic taxi service. Preliminary evidence from the expedition, Goldstein says, suggests this is exactly what's happening, though the degree of benefit and/or harm to various species will take years of study to work out.

Goldstein has an answer for those who might shrug and wonder if such questions really matter in the grand scheme of things. In a word: yes. And here's why: Half the oxygen we breathe emanates from microscopic phytoplankton sloshing around the surface of the ocean. After literally billions of years of performing that essential, priceless service, those vital organisms now must swim and feed and survive in a sea of plastic soup. Figuring out what's up with those organisms is, Goldstein suggests, a pretty vital matter. If we are inadvertently killing them off, the result could be far less visible, but even more devastating, than deforestation.

The other big questions that the SEAPLEX/Project Kaisei expedition sought to explore were equally compelling:

Now that we know that one in ten lantern fish has ingested plastic, what is this new part of the fishy food pyramid doing to these vital creatures that the rest of the food chain depends on? Many plastics can leach potentially toxic chemicals over time, particularly as the plastic begins to break down from the action of weather, wind and wave. Is that happening, and with what effect?

Are plastic particles acting as collectors of toxic chemicals, transporting and concentrating what is known as POPs—Persistent Organic Pollutants? This is the opposite of the leaching problem—plastics not giving off toxins, but acting as magnets for even worse chemicals. Pesticides, chemical fertilizers, half-combusted fuel, solvents and other man-made pollutants roll and rain into the oceans every day by accident and by design, and many of these chemicals are hydrophobic. That is, they hate water, won't dissolve in it and just wait for something better to come along that they can stick to. Weathered, cracked, sea-scoured bits of plastic become sponges for these POPs, and this is not a good thing. Yes, the plastic can take the chemicals from the water, but then little fish eat that plastic, and a chain reaction called bio-magnification begins.

This is the scenario the researchers are trying to gauge to see if it threatens marine ecosystems and human food safety: Let's say the little fish eats ten tiny pieces of POPs-infused plastic. Then a bigger fish comes along and eats ten of those tiny fish. Now we have a fish that has imbibed the equivalent of one hundred contaminated pieces of plastic. Then a bigger fish eats a bunch of those, and so on up the food chain, with the chemicals becoming progressively more concentrated in the larger sea creatures. This is bio-magnification. At some point, some of those larger creatures end up in the seafood case or the canned goods aisle at your local supermarket. We simply don't know what that means, but if Goldstein's team has its way, we will know in a few years.

"We just might not like what we learn," she says.

Even the most basic questions about the trash-ocean interface still await answers. The Scripps researchers are trying to accurately estimate the true size and concentration of the debris in the Pacific Garbage Patch and, more to the point, whether or not it is growing over time. The data is mixed on this: Observations in the Pacific by other researchers suggest the plastic has increased since the 1990s, even doubling in some areas of the patch. On the other hand, the largest collection of data from water samples in the North Atlantic gyre—twenty-two years' worth made by students on training voyages with the Sea Education Association—show that the plastic concentrations have held steady there, against all expectations. Researchers had assumed that, since plastic production has more than tripled in the past twenty-two years, there should be more plastic in the ocean, rather than the steady state it seems to have achieved (in the Atlantic, at least). Is there some mechanism removing the plastic—unknown currents, chemical reactions, plastic-eating microorganisms? Or is there really more plastic there despite the data, uncounted because it has broken down into such small particles that it remains undetected?

To help answer this, the Scripps researchers are doing two things: supplementing their manta tows with data from bucket samples from the gyre, and trying to figure out how plastic at sea ages. This latter problem is tougher than it sounds. Unlike archaeologists, who can carbon-date artifacts, or paleontologists, who can infer the age of a dinosaur bone from the geologic strata in which it was buried, the ocean plastics investigator has no way of telling the lineage of a 5-millimeter bit of plastic. The stuff has no chemical signature, no provenance, no forensic trail. It could be a year old, five years old, fifty years old—you just can't tell. In a landfill, you might infer the age of a piece of plastic from the junk it's buried with, the same as a geologist (*Oh, that hunk of blue polystyrene is sitting on top of a June 1973 edition of* Life *magazine—could be a clue!*). Ocean-borne plastic bits offer no such context. So Goldstein has pools of seawater filled with plastic baking in the sun and cooling at night on her lab roof back in San Diego, trying to come up with an age gauge for marine plastic debris. She'd like to run this experiment for two years or so; her professors told a stricken Goldstein they think it would be better to keep it going for two decades. Like landfills, she says, ocean plastic research is forever.

Everything about ocean trash is not a question, however. Here's what we do know: The United Nations estimates that a minimum of 7 million tons of trash ends up in the ocean each year, 5.6 million tons of which (80 percent) is plastic. The Sea Education Association data from the North Atlantic Gyre suggests that plastic concentrations in the ocean waters of the major gyres can easily reach 130,000 or more pieces per square mile of ocean surface; one survey of the Pacific Garbage Patch zone found concentrations nearly three times that level. The 5 Gyres research group, meanwhile, estimates that the total plastic content of the gyres exceeds (probably by a lot) 157 million tons, equal to 63 percent of all plastic made in the world in 2011. The group considers that estimate to be extremely conservative.

Even so, that's a big bag of plastic. It would take 630 oil supertankers to carry that much plastic. By contrast, the British Petroleum Gulf of Mexico oil spill in 2010, the largest maritime environmental disaster in history, released an estimated two-thirds of a million tons of crude oil. That whole oil spill could fit on two and a half supertankers.

To be clear, all of these ocean plastic numbers are at best educated guesses so far, based on slices of data collected from small sections of the biggest geographic feature humans will ever experience. There is a great deal of mystery left in that most ancient of things, the ocean, its newest resident, plastic, and how the two combine. A lot of the numbers and "facts" repeated in news coverage— claims that a hundred thousand marine mammals are killed each year from ocean plastic, that 80 percent of the trash at sea is from land sources rather than ships, that there is an actual garbage island looming somewhere in the Pacific—have no known sources with any credibility. The myth-making is a distraction, Goldstein worries, because the made-up information could erode the credibility of real science, and also probably understates the true problem. The research needed to firm up the data and answer the big questions is just barely getting under way.

As is often the case with environmental matters, more data often makes things look worse, not better. What we know so far makes clear that the matter of ocean trash goes way beyond what initially upsets most people: its aesthetics. It is yet another stress on a vital ecosystem that is already overtaxed by overfishing, acidification and climate change.

But plastics are a very different matter from global warming, about which politicians, if not many scientists, can find room to debate whether or not it exists and if it does, whether or not humans are causing it. As Goldstein points out, there's really nothing to debate about who and what is turning the oceans into plastic soup, as plastic is a completely man-made substance. It doesn't come from trees, volcanoes, space or bugs. It's all ours, and it enters the oceans through only one of three ways: accident, negligence or deliberate dumping.

"It's ours," Goldstein says. "We made it. We own it."

A HUNDRED years ago, not a shred of plastic could be found in the ocean because there was no plastic at all. It is hard to believe that the invention of the most ubiquitous substance in the human environment was preceded by radio, movies, recorded music, the airplane, the telephone, neon lights, air-conditioning, the lie detector, the electric vacuum cleaner, windshield wipers, color photography, the helicopter, the escalator, sonar, Kellogg's Corn Flakes and the theory of relativity. All were invented and put before consumers without need or benefit of plastic. All (except for relativity) are today unimaginable without plastic, from the helicopter's transparent cockpit bubble to movie DVD discs to the hose and power cord on the vacuum cleaner.

Plastic has gone so fast from zero to omnipresent that it's slipped beneath conscious perception. Take a moment and scan the room you're sitting in. Everything from pill bottles to DVD cases to the knobs on kitchen cupboards to the buttons on your pants to the elastic in your socks to the foam inside your seat cushion to the bowl you put your dog's dinner in to the composite fillings in your teeth— you get the picture—is plastic. It's everywhere.

Now take a walk on any public beach anywhere in the world and take a good, close look at the sand, at the broken bits of shells gleaming in the sunlight. Notice a flash of teal, a tinge of dark green, a bit of red or orange or yellow? Pick up the tiny sliver of color and see: Most are not pretty shell fragments, dried sea foam, eggshells or any other natural objects, though the sand and salt so readily camouflage them as such. They are bits of plastic, pieces of bottle

tops and cups, remnants of wrappers and foam cups. There is virtually no beach in the world where the sand is devoid of these synthetic particles, though the average beach walker rarely notices. You
have to really look closely for them. But when you do, the depress-
ing realization strikes: Once again, as with plastic in the home, they
are *everywhere*.

There is a special irony in plastic assuming the role of threat to
nature. Plastic conquered the world because, early on, the chemical
and manufacturing industries championed it as the miracle substance that would free humanity from the tyranny of nature. Piano
keys and billiard balls no longer had to be made from the ivory of
slaughtered elephants. Increasingly scarce metals mined across the
globe could be replaced by infinitely sculptable plastics that could
be produced in any half-decent laboratory in the country. Ladies'
stockings could be extruded from nylon-spewing nozzles instead of
silk-spinning caterpillars. All we needed were fossil fuels and imagination. Plastic was freedom.

The age of plastic (and the modern derivation of the word from
the ancient Greek *plastikos*, which means "moldable") started with
a Belgian-born American chemist, Leo Baekeland. He set up his
own research lab with the million dollars paid him by George East-
man, the father of popular photography and founder of Kodak, after
Baekeland invented a better type of photo paper. In 1905, the chemist used his Kodak earnings to finance experiments with a synthetic
form of shellac (a natural finish made from excretions of the female
lac bug found in India and Thailand). Instead, he stumbled on a
polymer made with coal tar and formaldehyde and a number of
inert ingredients (cornstarch among them) that could be shaped in
infinite ways, that dried hard and strong, and once set, proved highly
heat resistant—it wouldn't melt or lose its shape. He dubbed his

invention Bakelite, and it became the first completely synthetic in-dustrial and consumer plastic. It also is, to this day, the coolest plas-tic, rich, lustrous, solid and substantial in a way other plastics are not. The relatively heavy, durable, glossy Bakelite plastics were used in early twentieth-century telephones and radio cabinets—stylish retro items that are highly collectible today—as well as a host of more mundane items, from electrical insulators to chess pieces to cabinet knobs and Kodak cameras. Because of his invention's ver-satility, Baekeland chose for his company's emblem the letter "B" with the mathematical sign for infinity above it, which he had em-bossed on all genuine Bakelite products.

The success of Bakelite and the infinite possibilities it hinted at sparked a surge of experimentation and invention among the big chemical companies in the 1920s, 1930s and 1940s as they vied to patent the next "miracle material." In rapid succession, polyvinyl chloride (PVC, currently used in everything from plumbing to com-puter cases), Styrofoam, synthetic rubber and plastic wrap made their debuts in a variety of products, most of them commercial rather than aimed at consumers. A big exception to this marketing rule was nylon, the first synthetic plastic fiber, which was introduced to the public by the DuPont Company at an unintentionally appropri-ate location, the former massive landfill that became the site of the 1939 New York World's Fair. Initially developed as the ideal tooth-brush bristle, it was the formulation of nylon into synthetic silk that created a World's Fair sensation—stockings with no seams. More than 64 million pairs sold in their first year on the market.

As it did with other industries, World War II mobilization ramped up the plastics business tremendously. The First World War knew only wood, metal, wool, cotton and leather. A quarter century later, everything from combat helmet liners to parachutes, gun sights to cockpit windscreens was made from plastic, a quick and ready stand-in for scarce raw materials. The first iteration of Dow Chemical's Saran Wrap—which was a transparent green film with a putrid chemical smell—was used to wrap not sandwiches, but whole planes and artillery pieces to protect them from water and sea salt during transoceanic voyages.

When the war was over, plastic manufacturers had considerable excess capacity—and so new generations of products made of plastic were conceived, made and marketed whether they represented improvements over old materials or not. The disposable cups, spoons, forks, knives and plates that followed—an entire disposable economy—were born out of a kind of industrial hangover from the war effort, combined with cheap oil (the essential ingredient in many plastics) and America's then-ironclad control of the global oil supply. Now, though, plastic was pitched not as a substitute for the "real" thing, but as an improvement, a convenience, a freedom. Dow figured out how to make Saran Wrap clear and with no smell, and suddenly everything was being hermetically sealed. Plastic chairs, tables, counters, curtains and Tupperware invaded the American home (and, a short time later, the American landfill), supplanting wood, cloth, tile, metal and glass.

In the 1960s, plastic surpassed aluminum in volume as a raw material, and in the 1970s, it surpassed steel. It has continued to grow, reaching 51.5 million tons of plastic manufactured in 2010. That one year's worth of plastic outweighs the entire U.S. Navy's 286 active ships (which itself is so huge that the U.S. fleet represents more tonnage than the next thirteen largest navies of the world *combined*). Indeed, a year's worth of plastics would outweigh a navy of more than five hundred Nimitz-class aircraft carriers, the largest ship ever built, each one capable of carrying ninety aircraft plus more than five thousand crew and troops. Of course, there are only ten of these huge ships in existence. Plastic, when you hold it in your hand, seems so light as to be inconsequential, yet collectively it is that unimaginably huge.

This deceptive, alluring quality, plastic's horrifying convenience, helps explain why, for more than a half century, this miracle material, this great innovation that set us free, was only half baked, for no one thought through its life cycle, its afterlife. It takes 8 grams of oil to make a single plastic ketchup bottle, which will not be recycled because the ketchup residue inside is "contamination" and recyclers want clean plastic. Dirty plastic is just too hard to recycle, too costly. Failing at the birth of the age of plastic to think this through, to consider the life cycle of substances that do not occur in nature and that are, for all intents and purposes, immortal, is like failing to think through what to do with nuclear waste at the birth of nuclear power . . . which is exactly what we did.

Every year, a significant portion of this manufactured plastic remains unaccounted for. The American Chemistry Council reports that 34 percent of the annual plastic production, 17.5 million tons of it, is used for packaging—plastics that get thrown away very quickly. The EPA, meanwhile, has tracked 13 million tons of plastic packaging as waste, which means more than 4 million tons a year (the equivalent of forty of those super aircraft carriers) remain unaccounted for. Many ocean pollution researchers believe a substantial portion of this "ghost plastic"—these forty missing aircraft carriers we somehow misplace every year—finds its way into the ocean.

Which is how an oceanic garbage patch is born.

The most common types of plastic found there—primarily at the surface but also found as deep as one hundred meters—are polyethylene (used in a host of products, including plastic grocery bags), expanded styrene (Styrofoam), polypropylene (rope, nets, carpet, prescription bottles) and PET (notwithstanding its propensity to sink once broken up).

Whatever the type, pretty much every piece of plastic that ever entered the clutches of a gyre is still in there, ocean scientists say, except for what washes up on the beaches of Hawaii, which lie within the gyre's convergence zone and are inundated daily with plastic debris. They're still trying to come up with a number to describe the measurable and identifiable quantity of plastic in the garbage patches. Not supposition, not guesses, not extrapolations, but a real number.

So far, the scientific term most often used to describe how much debris is out there is: a lot. Some scientists, such as Miriam Goldstein, prefer a slightly more exact term:

"A *whole* lot."

So here's the big question, the one that eats at Mary Crowley and Miriam Goldstein and the crews who wander and plumb the five gyres: What sort of economy, what sort of society, could lose track of a fleet of forty aircraft super carriers of plastic year after year, without blink or blame?

And what, besides building oceans of plastic and mountains of garbage, can be done about it?

PART

2

THE TRASH DETECTIVES

An ocean of urban trash flows daily to my windy corner of San Francisco. Revelation blows in the wind: about the waste society, the careless or alienated urban dweller, environmentally thoughtless packaging and advertising, industry devoted to consuming without need.

—JO HANSON

Here's the main lesson of garbology: People forget, they cover, they kid themselves, they lie. But their trash always tells the truth.

—WILLIAM RATHJE

THE TRASH TRACKERS

WHAT IF YOUR TRASH COULD TALK?

What would happen if all the stuff we buy, use and ultimately throw away—a carton of milk, a computer keyboard, a case of beer, a color TV—could be aware of its surroundings and stream that information to us, forming a vast network of objects, a kind of Internet of stuff. Manufactured objects would become "blogjects," or "spimes," as science fiction novelist and futurist Bruce Sterling has dubbed them, so named because they would be aware of themselves in space and time, recording their own histories and travels, then passing on that information as it unfolds. Put aside for a

moment the scary, Orwellian, National Security Agency overlord
fears such a capability might represent, and consider just its implications for waste and wastefulness. Innate intelligence would, in
Sterling's vision, allow us to direct all objects—specifically the objects we throw away—to the best and most efficient path for reuse,
repurposing or recycling.

A world of "smart trash" would be a world in which zero waste
stopped being a distant dream and started being an achievable goal.

Of course, our trash isn't smart, and what we know about its
travels is shockingly thin at best, which is one reason the oceans
are slowly plasticizing. Retailers and manufacturers impose deep
scrutiny on the front end of the consumer economy, in which the
travels of goods on the way to market are compulsively microman-
aged, dated, accounted for and tracked with optical scans and RF
transmitters. That part of our consumer culture—the supply chain—
is brightly lit. But the *removal* chain, that's another story, a veritable
black hole. Not even the king of trash, the CEO of Waste Management, Inc., has a clear view of that dirty end of things. Which is why
the scientists and artists of the SENSEable City Lab at the Massachusetts Institute of Technology, inspired by Sterling's vision of an
Internet of things (with a generous dollop of funds from Waste
Management), decided to create smart trash. They gave select
pieces of trash their very own brains (i.e., the guts of a smartphone),
slipped them into the daily trash flow with the help of an army of
volunteers and a clever adaptation of global-positioning software,
then turned it all loose to see what would happen.

They are the nation's first true trash trackers. They followed
the meanderings of electronic waste to distant shores, of ratty old
sneakers that ran the equivalent of a dozen marathons aboard a
circuitous trash trail, and of printer ink cartridges that traversed the

continent not once but twice on the road to recycling—expending far more energy and resources in transport than could ever be recouped through recycling. Some materials intended for recycling simply never made it there.

Most studies of garbage are concerned with what's in our trash, and what happens to it once it gets where it's going. But the creators of smart trash wanted to expose *how* waste gets where it's going—the meandering, mysterious and, it turns out, occasionally disturbing path it takes after it is thrown away.

"Even the people working in waste removal don't really have a clear knowledge or picture of where the stuff goes," says one of the lead trash trackers, Dietmar Offenhuber. "We were fascinated to see an invisible infrastructure unfolding."

And seeing it, he says, is the first step in changing it.

SEATTLE RESIDENT Tim Pritchard learned about Trash Track in 2009 when he stumbled on a blurb on the Seattle Public Library website that said MIT scientists were looking for volunteers to participate in a novel experiment to "bug" trash with electronic trackers in order to better understand our waste. Intrigued, Pritchard immediately hit the e-mail link. This led to a correspondence with Offenhuber, and Pritchard soon was on board, becoming one of the more deeply involved of the several hundred volunteers who eventually participated.

At fifty years old, Pritchard was a natural for Trash Track. He'd been working to green himself for years, knocking his personal trash footprint way below the 102-ton legacy. He pegs his trash output at a single paper grocery bagful a month, recyclables included, though he qualifies this achievement by saying he's single and travels often for work, which cuts down his trips to the home trash can and recycling bin. He spent much of his professional career engineering audio for Broadway musicals—he toured with *Phantom of the Opera* for years—but more recently switched to corporate gigs, engineering the audio for conferences and major meetings, which are a form of theater unto themselves. He was fascinated by the opportunity to dig into the fate of trash, something he realized he knew next to nothing about, despite his personal dedication to sustainability.

Seattle was a good choice for the project's main push, too. Trash Track had launched smaller-scale efforts in London and New York, but those were primarily to create public exhibitions and museum displays featuring local "trash trajectories." The idea originated not as a research project but as a proposal for an exhibition sponsored by the New York Architectural League entitled "Toward the Sentient City." Cambridge University professor Rex Britter, a visiting fellow at the SENSEable City Lab, suggested that electronically tracking garbage in the city could produce some interesting revelations about waste for the exhibition. It soon became apparent that, in addition to the public education aspect that exhibitions provide, some real and original science could be accomplished, and the MIT group planned a larger-scale effort. And for that, Seattle was chosen, in part for its diverse transportation infrastructure. The city is a major port and interstate hub and a convergence for major rail lines, all of which figure in the waste "removal chain," including the daily, mile-long garbage train that hauls Seattle's trash to a landfill in eastern Oregon. The researchers also chose Seattle for its reputation as a green community that invests in sustainability. While the average U.S. city recycles about 30 percent of its waste, Seattle topped 50 percent the year Trash Track came to town. Seattle residents have also accepted a trash system that charges them by the amount of trash they produce rather than a straight monthly fee—a firm fiscal incentive to waste less. Offenhuber figured they'd have so many volunteers they'd have to turn some away, which they did.

Pritchard assumed the role of trail guide for the Massachusetts-based researchers. He showed them around town, identifying a diverse set of neighborhoods to assure the broadest possible spectrum of trash, income levels and lifestyles. And he chauffeured the Trash Track crew to volunteers' homes where they could root through trash and find the selection of items they wanted to track. Then he helped attach electronic tracking devices to the waste—making the trash smart. It was, Pritchard recalls, a blast.

The tracking technology boiled down to three parts: high-tech, low-tech and subterfuge.

The high-tech part required some compromises. Offenhuber and his crew wanted to deploy more than three thousand pieces of smart trash in order to have a sufficient sample size, so the devices had to be relatively inexpensive. They were going on a one-way trip, with no chance for retrieval. The smart trash trackers would meet one of three fates: they'd run out of power partway through their disposal journey; they'd be destroyed once they reached their final destination at a recycling center, landfill or hazardous-waste site; or they would simply disappear from view—out of range or out of luck.

The MIT scientists, after rejecting a variety of other gadgets and technologies as either unsuitable or too expensive, came up with a device that amounted to cannibalizing the innards of cellular phones and combining them with a custom circuit board and set of software instructions that turned phones into trash-tracking marvels. Once a trash tracker was placed in the waste stream (a public trash can, a recycling drop, a curbside bin), a motion sensor woke up each device whenever it moved, at which time the software commanded it to scan all channels and bands, locate the strongest dozen cell towers nearby, and store their identities with a time and date stamp. This information, which would allow the researchers to triangulate the smart trash's location at the time of the scan, was then compressed and sent as a simple text message to a server at MIT. The length of time the smart trash "phoned home" in this way was kept to an absolute minimum in order to preserve battery life. The devices went to sleep whenever they stopped moving as a further energy saver. Smart trash had enough battery life to last up to thirty hours of constant motion; standing still, it could last through three to six months of hibernation. This was important, because nobody knew at the outset just how long a piece of trash might travel or sit untouched before reaching its final destination.

The low-tech part of the tracking project was how to attach these transmitters to real pieces of trash. A variety of tapes, glues and other techniques, including latex rubber and carbon fiber, were tried and rejected either because they were too cumbersome or ineffective, or because they could not be applied quickly in the field by volunteers without advanced engineering degrees. They finally hit on inexpensive spray cans of epoxy foam used to patch holes in boat hulls. It could be quickly applied to just about any kind of trash, plastic, paper, metal or cloth, and dried fast and hard.

The subterfuge part involved hiding the tracking tags out of sight and out of harm's way—inside sneakers, nested in computer cases, wrapped inside old socks, tucked into café latte cups. This would protect the devices from being knocked loose by accident, or pulled out of place on purpose. Then the tagged items, the smart trash, had to be slipped inside the larger waste stream.

The smart trash was released into the wild in stages during the summer and fall of 2009. For stage one, a field test of the devices, Pritchard and the researchers drove around the city looking for abandoned trash and impromptu disposal locations. Along the way, they'd stumble on all sorts of trash just lying in the street or on sidewalks: old newspapers, soda bottles, paper cups—typical litter— and also discarded cell phones, dead or leaking batteries, abandoned washers, dryers and refrigerators, discarded computers, old furniture and car parts. A typical array of urban junk and flotsam was before them, stuff they'd normally drive by and barely notice, like weeds on an empty lot, except now, to the Trash Track team, these were golden opportunities. These random pieces of trash were tagged, then dropped into public waste cans, taken to recycling stations, slipped into homeowners' trash bins or just left on the street to see what would happen. This release validated the technology in the field and also let them perfect their curbside conversion of regular trash into smart trash.

In August 2009, stage two—the release of six hundred pieces of smart trash—began with a public event at the main branch of the Seattle Public Library, which served as Trash Track's home away from home. Volunteers were asked to come to the library with a piece of trash to be tagged and to sit for an interview. Then they were supposed to take their newly upgraded location-aware trash at least ten blocks away and dispose of it properly. Everything from old sneakers and T-shirts to broken electronics to milk cartons were brought in and tagged. Pritchard ended up tagging twenty items that he had been collecting—an old cell phone, a dish towel with a big hole in it, a junk drawer full of stuff he hadn't known what to do with. Sometimes the Trash Track team went to the volunteers' houses and tagged an array of household trash on the spot, then asked the volunteers to dispose of the items curbside as they normally would.

A month later, building on the experiences of the smaller tracker releases, Trash Track set loose a much more systematic and specific swarm of smart trash—2,200 pieces in all. One hundred homes were selected across the city, along with numerous elementary and high schools. This time, instead of relying on volunteers to choose the items of trash, the researchers had a wish list of items drawn from the EPA's trash categories, so that trackers could be attached to every type of waste imaginable: paper, cardboard, organics, leather, rubber, plastic, glass, metal, textiles and e-waste. The team particularly emphasized tagging "emerging" types of trash such as cell phones, fluorescent bulbs and other household hazardous trash that represent a relatively new and growing part of the daily American waste stream.

Weeks later, the volunteers and the public were invited back to the library to learn the early results of the project. Large color computer displays gave a real-time visualization of the journey each piece of trash had taken. Each piece of trash on the display had its photo posted, a written description, the location and time of disposal and an animation that showed its travels from the moment it was thrown away until it landed wherever it was heading, or its signal was lost. The display resembled something from the Pentagon during the Cold War—or at least a Hollywood depiction of it—in which the trajectories of incoming missiles were displayed. Except instead of warheads, the objects arcing across the country were pieces of refuse.

A runner saw her old sneaker had meandered 337 miles from Seattle to the Columbia Ridge Landfill in Arlington, Oregon. A plastic traffic cone made it 6.6 miles to a waste-transfer station, then vanished—its tracker either removed or destroyed, the fate of the cone an unknown. A coffee cup took more than seven days to traverse the city, picked up and dropped off multiple times. Cell phones ended up in Florida (Miami and Ocala), Ohio, Texas and a number of points in between. A cardboard box was driven a mere 3.3 miles to a recycling center. A lithium battery was trucked more than two thousand miles to Minnesota, while a printer ink cartridge was flown by Federal Express to Memphis, then driven 231 miles across the state to a recycling facility in La Vergne, Tennessee. Other ink cartridges went hundreds, sometimes thousands, of miles to other destinations; one first went east to Chicago, then returned west to Southern California.

Many electronic items went to ports where they were loaded aboard ships, then left cell coverage and were never heard from again. The distances that electronics and hazardous waste traveled were significantly longer than all other waste categories—the first time this was documented in a systematic way.

Trash Track, according to Offenhuber, has started to point out some major inefficiencies in the waste stream by bringing transparency to the normally invisible removal stream. It raises serious questions about the efficacy of current recycling efforts, which all too often send certain kinds of waste great distances, expending fuel and energy that could be conserved if more waste and recycling was handled locally. One example: There are only thirteen facilities in the world certified to recycle cathode-ray tubes—the now-outdated tube-style TVs and computer monitors that are still very common and are chock-full of hazardous materials. All thirteen of those facilities are in China. This is the sort of recycling that makes little sense as a strategy for sustainability.

The use of large numbers of volunteers transformed the project into a kind of "citizen science" effort, Offenhuber says, and that gave it unexpected power. People got invested. They came to the library to follow the course of a sneaker or a cell phone. Schools were excited at the classroom discussions the project engendered. There was even talk of developing a kind of "Trash Track in a Box" for educational use—a self-contained, prepackaged Trash Track kit. A number of companies contacted MIT about replicating the experiment to track their own waste, and the technology company Qualcomm, which worked on the last generation of trackers for the project, has come out with a commercial version of the device for sale to businesses and communities interested in exploring their own removal chain.

Offenhuber thinks anything that gets ordinary citizens involved in understanding and bringing transparency to the fate of their trash represents an important step forward. He sees urban planning in the twentieth century as dominated by a paradigm he calls "infrastructure as an invisible black box," which not only keeps regular people in the dark, but leaves them feeling helpless about doing anything to make it better. A twenty-first century of smart trash, an Internet of things, can turn that around, he says. Each trash trajectory arcing across the country smashes the invisibility that has long masked our trash and its disposal—along with the illusion that our trash is handled efficiently.

Recycling in particular has long served as a balm and a penance—a way of making it okay to waste, the assumption being that if something is recycled, then the energy and materials are not being lost, and our disposable economy of abundance doesn't really seem so wasteful after all. But the meandering, inefficient and sometimes purposeless paths for our garbage revealed by

Trash Track puts the lie to those old assumptions. There is no penance for being profligate when the waste-management system itself can be so unpredictable and, at times, incredibly wasteful. When a printer ink cartridge can make multiple transcontinental trips before finding its way to a recycler, it creates a footprint that's more environmental disaster than savior.

The meandering maps and trajectories revealed by Trash Track have provided part of the answer to the second big waste question that must be answered in order to wrestle the 102-ton legacy into submission: How is it possible for people to create so much waste without intending or realizing it?

For one thing, it seems, our waste doesn't go where we think it goes. We aren't counting, mapping or directing it well. The idea that there is a waste-management "system," it seems, is more illusion than reality. At best there is a chaotic hodgepodge of potential trash

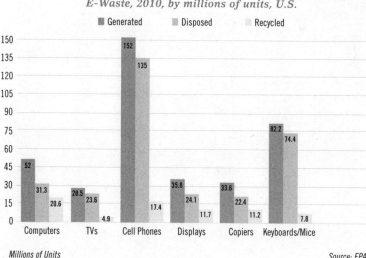

E-Waste, 2010, by millions of units, U.S.

Generated   Disposed   Recycled

	Computers	TVs	Cell Phones	Displays	Copiers	Keyboards/Mice
Generated	52	28.5	152	35.8	33.6	82.2
Disposed	31.3	23.6	135	24.1	22.4	74.4
Recycled	20.6	4.9	17.4	11.7	11.2	7.8

Millions of Units   Source: EPA

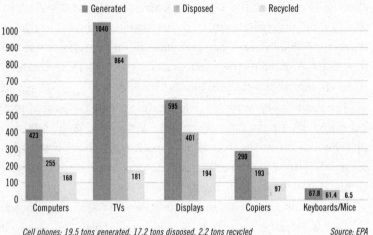

E-Waste, 2010, by thousands of tons, U.S.

■ Generated   ■ Disposed   ■ Recycled

Cell phones: 19.5 tons generated, 17.2 tons disposed, 2.2 tons recycled   *Source: EPA*

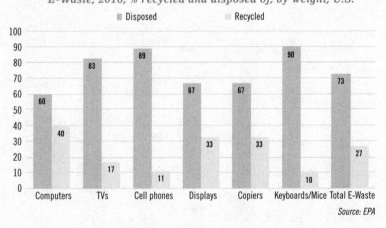

E-Waste, 2010, % recycled and disposed of, by weight, U.S.

■ Disposed   ■ Recycled

Source: EPA

destinations that eludes both control and detection in ways that would never be tolerated in other industries and supply chains. This revelation suggests that the second big question should be modified slightly: How can we ever put an end to waste if we can't even keep track of it?

Smart trash provides an inkling of the power over waste that could be achieved with a little more garbage brainpower.

PERHAPS THIS future is on its way. The SENSEable City Lab is trying to scale back its scattershot attack on trash by narrowing the focus in a follow-up project called "Backtalk." This time the tracking is aimed at one specific type of trash: e-waste. The goal is to find ways to shorten the journeys this growing and often hazardous type of electronic trash is taking, and to examine just how much of it ends up exported to toxic salvage yards abroad. Estimates run from 20 to 80 percent of U.S. e-waste gets offshored for disposal, with dire health and environmental consequences on the receiving end.

"Watching the path our trash takes was fascinating and surprising," says Tim Pritchard, "and sometimes disappointing. Seattle is a community that's made a lot of progress on sustainability, but this has shown us how far we have to go . . . The clock is ticking. If we don't embrace a different way, as awkward as it may seem, there will be fairly dire consequences."

8 DECADENCE NOW

THE BUCKET AUGER CHEWED DEEP INTO THE ground, a three-foot-wide steel cylinder with three-inch jagged teeth bristling from its business end. The "bucket" part of the bucket auger is a yawning, spinning maw that grinds through earth as if it were made of marshmallow. Mounted on a telescoping pole, it's capable of burrowing a hundred feet down, then retracting and bringing up huge chunks of whatever lies beneath, bucketful by bucketful, a whirling sand toy on steroids. The bucket auger's torque is so powerful that it has chewed right through a wrecked and buried car, engine block and all.

Pulled up and upended with a hiss of hydraulics and grinding of gears, the bucket disgorged a blend of dirt and plastic and old newspapers, many of them yellow and brittle but surprisingly readable. There were cans, yard clippings and several hot dogs, a bit dingy but intact—the queasy power of preservatives at work, perhaps. And there was a white ceramic bowl of some brown stuff which, when its dirty crust was scraped away, revealed something bright green inside. There were chunks of something still visible in the mix.

"Hey! I think it's guacamole!" archaeologist Bill Rathje shouted to his crew of student volunteers, dabbing his finger in the stuff.

It was guacamole. The chunks were avocado slices, still green. And the nearby newspapers allowed the perishable treat's age to be inferred: twenty-five years. The guacamole had last seen daylight a quarter century ago and yet, while not exactly edible, it seemed fresher in appearance than it would have looked after just a few days sitting in Rathje's kitchen sink. It had been preserved—unintentionally and, to all but Rathje and his crew, unexpectedly.

Because that's not how landfills were supposed to work. Or so it was said.

The bucket augur is a tool for drilling wells, for water and oil, but it's also how Rathje, founder of the Garbage Project, spent decades exploring the inner space of landfills, about which many knowing assumptions have been made over the years—and which Rathje, time and again, proved mistaken.

He is the world's first garbologist, and his work uncovered just how poor an understanding we have of our own waste.

"Most people don't really know their trash," says Rathje, a broad, deep-voiced archaeologist who has been labeled the Indiana Jones of refuse. "But through their trash, we sure do know a lot about them."

THE IDEA for a Garbage Project—for a systematic and unprecedented deep analysis of modern waste using the same skills, tools and modes of inquiry archaeologists employ to understand the ancient world—began with a simple student project and training exercise in the early seventies at the University of Arizona in Tucson.

Rathje, a Harvard University–educated archaeologist who specialized in the study of ancient Mayan ruins, was then a young professor on the Tucson campus, long a hotbed of archaeological discovery due to the wealth of ancient Native American sites throughout the region. He wanted to introduce basic archaeological methods to the students in his anthropology class through a series of independent study projects. Two of the students came up with the idea of fact-checking some typical cultural stereotypes with physical reality—which they proposed to accomplish by comparing the real-world trash collected from two homes in an affluent area of Tucson with the trash from two homes in a poor part of town. Would the two sets of families differ in unexpected ways? Or would they be unexpectedly similar? Would the real-world detritus produced by the test subjects (their identities protected by the archaeologists' dusty equivalent of doctor-patient confidentiality) confirm cultural clichés, or shatter them?

This idea appealed to Rathje on a number of levels. For one thing, he's a natural contrarian, so the idea of using trash to upturn stereotypes and commonly held assumptions was beyond delicious. "Cut the crap!" and "Bullshit!" are favored expressions of his, reserved for what he considers to be galling misstatements about garbage by the uninitiated. (Over the years he has been particularly incensed by persistent claims and extensive media coverage of the supposed evils of disposable diapers, which he says create "barely a blip" in the average landfill, while distracting the public from genuine and larger garbage problems, such as the proliferation of phone books, most of which are unwanted and, to this day, mostly get landfilled instead of recycled.)

The other thing Rathje liked about the student garbage study was its embrace of the gritty realities of genuine archaeology, which, for all its seeming romance, its air of exotic locales and lost civilizations, really boils down to rooting around in dead people's trash. Really, really old trash, certainly, long stripped of its smells and general ickiness, but trash just the same, the true object of archaeologist lust because it represents the unvarnished story. The monuments, stone tablets, formal histories and burial chambers that describe the glories of dead civilizations are all well and good, but they tell the story that the kings and scholars wish to communicate, the idealized version, the version that the victors in a war get to tell rather than the stories of the conquered. In garbage, though, there are no half-truths, no spin, no politics. Conquerors may plunder the riches and thereby the historical record, but no one plunders trash. The accrual of what a people ate will be there, master and slave, worker and lord alike, an honest tale of crusts, rinds, bones and seeds. How they lived, what they wore, where their trade routes reached, even how and who they worshipped—all of that, and so much more, is contained in the record of their garbage, the unbiased arbiter and keeper of the inner life of any civilization. What we say about ourselves, observes Rathje, is never as honest or as revealing as what we throw away. This is why archaeologists crave trash, why Washington's outhouse at Valley Forge was a major dig (the general and future president threw all kinds of things in there—then trash, now illuminating artifacts), and why it was not such a stretch for a professor of ancient Mayan culture to approve his students' plans to look for similar truths in Tucson's trash.

Based on the garbage recovered, the two students concluded that, income (and conventional wisdom) notwithstanding, the two sets of families consumed similar amounts of steak, hamburger and milk. The poor families, however, bought more household cleansers and spent more on goods related to child education. As fascinating as these differences were, the tiny sample size made it impossible to draw any sweeping conclusions. But it did suggest a new, potentially fruitful subject of study: using trash to gauge all sorts of contemporary behaviors, and to see if that trash trail squared with our societal assumptions, or revealed the myths we live by.

This was new territory. Detectives and journalists had been known to root through garbage from time to time, looking for stories and scandals in those pre–paper shredder days. But a scientific inquiry into the patterns and context of trash as real-time cultural artifacts, evidence of consumer behavior and window onto society's soul had never been attempted in any sustained way. During World War II, the Army tasked a pair of enlisted men with marketing experience to gauge soldiers' satisfaction (or lack thereof) with military mess by analyzing the food that was thrown away by mess halls. The results: too much food was being prepared in mess tents throughout the Army; staggered mess calls resulted in more clean plates than single, long lines that allowed the food to get cold for many soldiers; most of the soup, kale and spinach got trashed; and there was no such thing as too much ice cream. Menus and meal preparation were soon adjusted (less spinach and kale), and the Army began saving 2.5 million pounds of wasted food a day—the first modern practical benefits of the study of garbage or, as it has come to be known, garbology. Despite the rousing success, the Army discontinued its study of food waste (and garbage in general), and the thread wouldn't be picked up again until Rathje spotted the opportunity three decades later in Tucson. "We were," Rathje recalls, apologizing with unconvincing sincerity for the pun, "breaking new ground." In 1973, Rathje, several of his colleagues and his students expanded the garbage-study concept and the amount of trash to be analyzed, and the Garbage Project was born.

The conceit was simple: If we use the same archaeological tools and techniques previously employed on Egyptian pyramids, lava-encrusted Pompeii and the painted caves of Lascaux, what can we learn about American civilization from its garbage? What is the secret story of trash?

The archaeological team did not go out on "digs" at the beginning. Instead, the "artifacts" were delivered to the Garbage Project, which is to say, the university arranged to have the city sanitation department dump piles of garbage from specific census tracts on a campus maintenance yard six days a week. Then Rathje and his team of student volunteers surveyed and cataloged the mess, wearing rubber gloves, surgical masks and gowns, bagging and tagging the garbage on sorting tables, trying to figure out how to categorize a marshmallow. (Answer: as "candy.")

Before they could figure out what it all meant, though, they had to develop from scratch an entirely new language of trash. They went so far as to create a sort of Rosetta stone of aluminum can pull-tabs (this was back in the day when the tabs were designed to detach from beverage cans). It turned out there was a surprising variety in these little bits of metal that could identify beverage type, age and manufacturer simply by the shape and heft of the tab, and the Garbage Project remains to this day the one and only forensic authority on the subject. Meanwhile, an entire numbering system evolved over time to catalog the rest of the garbage: 001 was beef, 003 was chicken (the Garbage Project's nemesis, for nothing smelled worse than rancid uncooked chicken), 090 was powdered baby formula, 139 was a plastic container, 149 was auto supplies. There were 190 separate codes in all.

After the garbage was categorized, counted and compared, the unexpected and counterintuitive findings began almost immediately.

First, there was the matter of food waste, a major component of everyday trash. Food waste was rampant, though that wasn't news. What was surprising was that the amount of waste seemed to rise during times of shortages and high costs. This was particularly easy to spot when it came to meat, which gets trashed with unusually good documentation along for the ride—the meat packaging used at markets includes the type of meat, its cost, its packaging date and its weight. Comparing that to the actual meat discarded in the same batches of trash provides a reasonably accurate measurement of carnivorous food waste.

In that first year of the Garbage Project, a blight decimated feed crops, which drove up the cost of raising beef cattle, which in turn caused a sharp and well-publicized increase in the cost of red meat to consumers. In some areas, there was a shortage of popular cuts of beef, amid a great deal of media coverage about the turmoil in the beef cattle industry.

Common sense might suggest that such scarcity, high cost and feverish press would lead to a reduction in food waste, as families sought to stretch their food dollars and get every meal they could from each costly purchase. But the opposite was true. Beef waste during normal times hovered around 3 percent, the Garbage Project volunteers found. But during the shortage months, wastage tripled to 9 percent.

Rathje eventually hit on the explanation for this counterintuitive behavior. When shortages occurred (or were even discussed in the news media), consumers purchased more beef than normal. As hoarding exhausted supplies, they also tended to purchase cuts of meat that they normally did not buy and that they did not necessarily know how to prepare well. The combination of overbuying and bad cooking led to extra waste, with more raw, spoiled meat and more uneaten (and apparently unappetizing) cooked meat showing up in the trash than was the case during normal market conditions. Consumers, when asked, thought they were being sensible and economical, when their trash told a different truth: They were being more wasteful than ever.

In a similar vein, the Garbage Project discovered that well-publicized special collection days that sanitation departments set for collecting hazardous household waste—varnishes, paints, cleaning compounds, old motor oil, oven cleaners and other nasty chemicals that are not supposed to go in ordinary landfills, yet often do—had the unintended effect of leading to more, rather than less, improper disposal of toxics. City sanitation departments have in modern times labored to keep these toxic home products out of regular garbage landfills because of the environmental hazards they pose, which is why special collection days and locations are set for them. The Garbage Project analysts, who wanted to examine the effect of these toxic collection days, found that on the day after these special hazardous waste pickups, the regular trash stream had twice as much hazardous waste improperly tucked inside it as normal.

The explanation was simple enough: Alerted by the publicity about the hazards of such materials, people rounded up all those nasty cans and bottles of sludge and dried paint that had sat forgotten, gathering dust in their garages, cellars and sheds. Then for one reason or another, they had missed the special collection time. Chagrined but also motivated by the publicity to get rid of the stuff, they had just tossed it in their regular trash bins and covered it with orange peels and plastic debris. Once again, the Garbage Project had shown that a well-meaning trash policy based on assumptions about human behavior had generated the opposite result as was intended. Instead of cleaning up toxins, the special collection days were making things worse. Rathje suggested the best way to avoid future disasters would be to make many more frequent toxic pickups, or create a dedicated drop-off site that the public could easily access as needed.

An interesting Garbage Project aside: The trash from poorer neighborhoods could readily be identified by their hazardous materials, which were dominated by car care items, oils and additives; the toxics most common to middle-class neighborhoods were weighted toward paints, stains and varnishes—the substances related to home improvement; affluent neighborhoods, apparently focused on lawn care, had toxic trash dominated by pesticides, fertilizers and weed killers. The project developed a surprisingly accurate formula for calculating the relative income and demographics based on these kinds of trash distinctions.

Rathje also noted that when sanitation departments provide larger trash cans to households, those households immediately begin to produce more trash. He calls this Parkinson's Law of Garbage. It's based on the original Parkinson's Law, formulated by British bureaucrat C. Northcote Parkinson, who in 1957 noted that work expands in order to fill whatever time is available for its completion. The trash version of this principle holds that "garbage expands so as to fill the receptacles available for its containment." Rathje discovered this after researchers were perplexed by the fact that households in Phoenix threw away a third more trash than their counterparts in Tucson, despite the largely similar demographics, culture and geography shared by the two cities. At the time, Phoenix collected its trash with mechanized garbage trucks and 90-gallon standard bins; Tucson had smaller trash receptacles. In 1988, Tucson switched to the same system Phoenix used, and the average amount of garbage produced—a figure that had barely budged for fifteen years—suddenly went up by a third. The difference was made up by more yard wastes (had they been composted or just left on the ground before?); old clothes (had they been donated or given to others in the past?); household toxics (long accumulated in basements and garages); and recyclable plastics, glass and cans (previously bundled for separate collection, now quickly and easily dumped in the bigger bin). Parkinson's Law suggested the need for separate mechanized bins for recyclables, which has since become the industry standard.

Other garbage insights large and small emerged:

- Discarded birth control pill dispensers showed that a substantial minority of women were taking the pills incorrectly (missing and skipping days).
- The presence of condom wrappers in the trash rose 45 percent in the first two years after AIDS hit the news, suggesting that the public had taken seriously health admonitions to practice safe sex.
- Families in low-income neighborhoods tended to buy the smallest-sized packages of food, while the trash from affluent neighborhoods was rife with large- and economy-sized products—which means the poor end up paying more money

for packaging than food, while the food dollars of families with cash to spare go much further.

- The amount of alcohol consumed did not vary with phases of the moon, as legend has it, but drinking rates did increase considerably at certain times of the month: immediately after the paydays of major local employers.
- Finally, the Garbage Project issued a mild warning to romantics to rethink how they celebrate February 14: While almost no Halloween candy is ever thrown in the trash (only wrappers), a great deal of Valentine's Day candy never leaves the wrapper or box, and ends up at the dump instead.

These sorts of insights, whether they suggested that serious policy changes were in order or merely served as fascinating trivia, had a cumulative impact: The reputation of the Garbage Project, which began as something of an oddity that newspapers and local TV broadcasts delighted in treating tongue-in-cheek, and that initially was a source of embarrassment in the academic community, gradually was transformed. Yes, the word "garbology" may have originated as a sort of joke, first used in the 1960s by municipal dustmen in New Zealand and Britain to make their job title sound loftier. But that began to change. The next edition of the *Oxford English Dictionary* defined "garbology" as Rathje did: "the study of a community or culture by analyzing its waste." Rathje's papers on trash were being accepted at major scientific journals. The Smithsonian Institution wanted to put together a garbology exhibit. This wasn't a joke after all—there was real science to be done here, and real revelations coming out of it.

Then Hollywood got in on the act, having discovered one of Rathje's first garbage sorters, a student named Sheli Smith, who worked six semesters with the Garbage Project. Smith, now with the Past Foundation in Ohio, went on to an illustrious archaeological career of her own, including the 1982 excavation of a colonial sailing vessel from 1710 that was found ten feet below the surface of Water Street in Manhattan's financial district, which—you guessed it—is in large measure built on landfill consisting mainly of eighteenth-century garbage. Smith was invited to appear on a popular television game show called *What's My Line?*, in which people with unusual careers attempted to stump a panel of celebrities. Smith won, having cleverly gotten a manicure and an orangey fake tan atop her real desert tan just before the show; the celebrity judges concluded that no one with nails that gorgeous could possibly be a professional garbage sorter.

Next thing Rathje knew, representatives of the U.S. Census were on the horn, looking for help. They were having a terrible time trying to calculate the number of households in poor communities, and in particular, the number of two-parent households. Census leaders were smarting at the revelation that they had a 40 percent margin of error when it came to determining whether or not there was a father present in inner-city and immigrant neighborhoods. This, it turned out, was a crucial question, and not just as a matter of academic interest. These figures would determine all sorts of government policies, from the shape of state and federal voting districts, to the amount of child welfare payments allocated (which is why the Census was having trouble: some residents feared being counted and losing benefits they relied on to live), to the budgeting and placement of social services for schools, daycare, single moms and needy kids. Forty percent error rates just would not cut it. Could the Garbage Project help? Could Rathje's unusual insights into trash demographics be used to determine the age and gender of residents in a given neighborhood based on what they throw away?

It turned out that they could. The Garbage Project had studied in detail the food and garbage patterns of two hundred households for five weeks, painstakingly sorting and weighing all their trash, then subtracting yard waste because it varied too much between urban, suburban and rural locations and so could skew the results. These households were active participants in the study, answering extensive questionnaires, so the Garbage Project knew the exact population, gender and ages of all the family members involved. Rathje was then able to construct an equation: x households multiplied by y residents equals z pounds of garbage. As long as you knew the value of two of those numbers, you could figure out the third. The Garbage Project had produced what Rathje called the "magic number" to plug into a population equation. Multiplying that magic number by the number of households in a given neighborhood would tell you with surprising accuracy how many people lived there. This held true across geographic regions and income levels. Subsequent tests of the equation, according to Rathje, showed it had an accuracy of plus or minus 2.5 percent, which was better than the Census Bureau had managed in many areas of the country.

Figuring out the second part the Census wanted—gender—turned out to be more difficult, however. This is because there are few distinctly "male" pieces of trash—both men and women can use the same sorts of razors and shaving cream, for instance, and not even the presence of male contraceptives would establish actual residency. Those items that are exclusively male, or close enough—men's underwear, or cigar butts, for example—occur with such infrequency in the waste stream as to be useless as data points. The researchers got around this in the end by figuring out the markers for everyone but men. They developed equations for the number of infants (diaper counts), the number of children (discarded toys, toy packaging, children's clothes and their packaging) and women (discarded female-hygiene products, cosmetics and women's apparel), from which they could extrapolate the number of men from the total neighborhood population count. The mystery of the uncounted fathers residing in some neighborhoods could at last be solved, and the Census could cure its chronic undercount. The plan was to apply this new technique in time for the 1990 Census.

But it never happened, Rathje lamented. The then-director of the Census's Center for Survey Methods Research decided that it would be bad public relations to hire someone to analyze people's trash. They'd just have to live with the undercount.

Nevertheless, this work showed a new and more powerful side of the Garbage Project, as it moved beyond simply sorting trash and into comparing its real-world footprint with the results of surveys and polls. It became very clear that trash provided potent, unique clues about the inner working of society and country that could be found nowhere else. It also began to show why trash was such a social, environmental and fiscal problem: Most people had no idea what was really in their garbage (or, for that matter, in their closets, refrigerators, cupboards and shopping carts).

The Garbage Project was tackling a big piece of the second question that must be answered in order to shrink the 102-ton legacy, namely why we are also so obviously clueless about the true size and nature of our waste. Rathje was exposing our trash mythology: what we know versus what we think we know about garbage.

Rathje and his students soon documented how average Americans overestimated their intake of healthy foods, claiming, for example, to eat three times as much cottage cheese as they actually

purchased (based on the number of containers found in their garbage). And they vastly underestimated their less healthy eating
habits. Potato chips, for instance, were reported to be eaten in quantities 81 percent smaller than the crumpled chip bags in the trash
actually documented. Rathje called this the "Lean Cuisine syndrome." This kind of data is psychologically unsurprising, as most
people chronically overestimate their "good" habits and underestimate the "bad." But it also suggested that the focus groups and consumer preference surveys that so many business decisions are
based on are practically worthless.

Alcohol consumption was among the most dramatic deviations
between survey and trash can, with a vast disparity between what
people claimed to have imbibed, and what the empties in the waste-
basket indicated. About three-quarters of households reported zero
alcoholic beverage intake during a typical week, while 20 percent
reported seven or fewer beers consumed, with a handful owning up
to drinking more than that. The trash reality check turned all this
on its head: Only one-quarter of households had no evidence of
alcohol in its week's worth of trash. Another quarter showed one
to seven beers consumed. And fully half the households had
consumed eight or more beers in a week. And this was after the
garbage sorters excluded the debris from data-skewing parties (dis-
cerned by the presence of soggy paper plates, large numbers of
disposable cups and the telltale presence of cigarette butts in partially empty beer bottles).

Interestingly, while people tend to underestimate their own
drinking by 40 to 60 percent, in households where one or more
adults are teetotalers, they tend to be uncannily accurate in their
estimates of the drinking habits of other family members, within an
error rate of 10 percent or less. Rathje called this the "surrogate

TRASHY DELUSIONS

From a Garbage Project study for the U.S. Department of
Agriculture, on the Lean Cuisine syndrome (how people
overestimate and underestimate their consumption of certain
foods based on whether they are fattening or not):

% Underreported		% Overreported	
Sugar	94	Cottage cheese	311
Chips/popcorn	81	Liver	200
Candy	80	Tuna	184
Bacon	80	Vegetable soup	94
Ice cream	63	Skim milk	57
Ham/lunch meats	57	High-fiber cereal	55

syndrome," although others have suggested "town gossip complex"
might more accurately describe this phenomenon.

As far as the type of alcoholic beverages consumed, the Garbage
Project found a broad disparity across neighborhoods and income
levels. The alcohol-related trash from low-income areas was dominated by beer bottles, with a smattering of hard-liquor containers
mixed in. Middle-income neighborhoods had booze-related trash
that spanned the entire spectrum of spirits: beer (mostly in cans),
wine and liquor. Upper-income households showed more expensive
wine bottles than their middle-income counterparts, but somewhat
less prestigious hard-liquor brands.

None of this is very surprising, as Rathje saw it. The interesting
part is that eighteen years of data show that the actual alcohol content delivered by these various beverage choices is consistent

across all income groups. Regardless of income, Rathje found, everyone on average gets the same buzz on.

ONE DAY, after some sixteen years of trash sorting and household consumer surveys, Rathje and a colleague were discussing the Garbage Project's latest findings. The other archaeologist heartily congratulated Rathje for all the fine work, but then made a pithy observation. "That's great and all, but where's the dirt, Bill? If there's no dirt, it's not archaeology."

Rathje was brought up short by this. His colleague was right: Archaeologists dig. If they wanted to do real archaeology, garbologists would have to dig, too. Why hadn't he thought of this before? It was time to stop bringing the garbage home, and start bringing their project to the garbage.

Thus began years of plumbing the depths of landfills—twenty-one of them, all over the country, more than 130 tons pulled up by the bucket augur before Rathje finally called it quits after more than thirty years as the world's leading garbologist.

The single most startling finding from Rathje's excavations was that garbage does not decompose inside landfills as most people, including sanitation experts, believed. A well-maintained, airtight, dry sanitary landfill was more like a mummifier of trash than a decomposer of trash, Rathje found. Fifty-year-old newspaper was intact and readable, headlines about President Truman's electoral chances still bold and black on the front page. Steaks and hot dogs came up intact after decades. (But kaiser rolls? Not so much: Exhumed, they looked remarkably like ancient, mossy granite grinding disks used to make prehistoric cornmeal. Then Rathje spotted the poppy seeds and realized he had not fallen through some weird trash time warp that put Stone Age tools in the same landfill stratum as bottle caps and an exhausted tube of hemorrhoid cream.) Landfills, the Garbage Project diggers proved, were in many ways like giant time capsules, preserving for decades the seemingly perishable items we expected would turn to organic mush, while other items very, very slowly decomposed. There's enough decomposition to generate a steady flow of methane, but at a slow enough rate that organic waste remains recognizable for a long time—grass clippings still green after fifteen years, onion peels and carrot tops hanging in there after twenty—which means that the methane flow can continue for a very long time, too.

According to Rathje, these findings, while unsettling to the orthodoxy, are a good thing. It means some of the potentially toxic juices people feared would leach out of landfills are basically just sitting there. This stability had long been recognized as the silver lining of plastic trash that we fail to recycle–it didn't decompose, and so posed no environmental hazard as long as it was contained in a landfill. On the other hand, the materials that people had hoped would biodegrade—even the stuff officially designated as (or specifically designed to be) biodegradable—didn't break down as expected in landfills, either.

There was a bad-news, good-news finding on hazardous waste in municipal landfills, too. The bad news: There was a lot more of it than anyone had believed. There were twice as many cans of bug spray, containers of paint and old drain-cleaner cans being slipped into trash bins and spirited off to landfills as had been believed. The good news: Like so much other stuff in the landfill, it mostly just sat there. Even when the containers leaked or broke, the surrounding "trash matrix" soaked it up like a sponge and retained it. And a little more bad news: When there was a problem of landfill contamination leaking out into the real world, particularly after floods, this presence of chemical hazards could make residential trash just as toxic as industrial waste.

The other dramatic finding from the landfill excavations, one that shocked even the jaded garbage sorters from Tucson who thought they had seen it all, was the amount of food waste dumped in landfills. As much as 17 percent of the garbage by weight that they were hauling up in the late 1990s and early 2000s consisted of food waste. Some of it was truly waste—coffee grounds, eggshells, plate-scraping slop—but nearly equal portions were completely edible, from expired hamburger to potato peels (a major and completely edible weight component of food waste) to those specialty breads such as those deceptive kaiser rolls, which ended up landfilled at far greater rates than standard loaves of bread, which were practically no-shows. Indeed, that finding led to the Garbage Project's "First Principle of Food Waste":

The more repetitive your diet—the more you eat the same things day after day—the less food you waste.

This principle upsets quite a few people and special interests, Rathje soon discovered. Nutritionists want a healthy variety. Food companies live and die by novelty, constantly introducing new breakfast cereal variations and reformulated baked goods and new flavors of processed food. But novelty (which consumers think they want more than they *actually* want it) breeds waste—those darn kaiser rolls, along with hot dog buns and biscuits and English muffins, end up getting thrown out anywhere from 30 to 60 percent of the time. Novelty may make for effective marketing, but in terms of waste, it's a disaster.

America's propensity for throwing away perfectly good food that could quite literally end hunger for millions of people has received considerable attention (if not reform) recently, but the Garbage

Project was calling attention to food waste as a vital issue fifteen years ahead of the curve.

"We just thought it was appalling," Rathje recalled. "And most people are oblivious to it. If you ask them, they'll tell you they are careful not to waste food. But as usual, their garbage tells a different story. It was typical for the households we looked at to waste 15 percent of the food they bought."

A number of landfill excavations were made through contracts with cities that needed better insight into their trash. Unearthing garbage in Phoenix, the researchers were able to determine the amount of recyclables that were being buried. Aluminum cans alone could net the city more than $6 million a year if captured, recycled and sold at market rates. The city public works department used Rathje's analysis to pry $12 million from the Phoenix city council to launch a new recycling program for the Arizona capital.

After the Garbage Project informed the city of Toronto that construction waste was clogging a fifth of their available landfill space, the city invested in the infrastructure necessary to recycle concrete, bricks and other demolition and construction debris. Excavations of four landfills in Toronto also validated the city's recycling program, one of the oldest in North America, which was under fire for costs. Rathje and his crew proved it was biting deeply into the waste stream and, if anything, had surpassed expectations. And in Mexico, the government adjusted its import taxes in favor of a bit of protectionism when the Garbage Project found that luxury goods purchased in Mexico City's affluent neighborhoods tended to be American-made.

RATHJE MADE an estimate a few years back that suggested all of the garbage produced by the United States for the next thousand years could fit inside a single landfill—as long as said landfill stretched across forty-four square miles and rose 120 feet high.

That sounds huge, but not as huge as most people think all the country's trash should be. Such a landfill (less than a quarter the height of Puente Hills) would cover all of the Bronx, or a mere one-fifth of the West Coast's main Marine Corps base, Camp Pendleton, or just .036 percent of the land area of the state of New Mexico. More square miles of that state's national forests have burned in a single fire season than such a landfill would cover in a thousand years. In other words, a thousand-year landfill would be big, sure, but not really all *that* big. No one is proposing such a mega-dump. The point is, Rathje liked to say, we have plenty of room to keep burying our trash until we find a better plan. Space for trash, in other words, is not the problem.

Of far greater concern, as Rathje saw it, is the trash that *doesn't* get into the landfill vault—the debris in the gulches, the plastics in the ocean, the waste that drifts off into rivers and streams. And the biggest system flaw of all, he argued, is the disposable, wasteful mind-set that creates the flow of trash in the first place. Rather than a problem specific to landfills or other sanitation strategies, Rathje always maintained, this is a flaw in how manufacturers create and consumers use disposable products.

Rathje retired from the garbology business in his early sixties and spent the last years of his life devoted to Buddhism and his passion for photography, illustrating Buddhist texts with his photos of nature. In what would be his last interview, Rathje shared his thoughts about the state of garbage in America for *Garbology*. He died of natural causes a few months later in May 2011, at age sixty-six.

He confessed to being more than a little disappointed that the

Garbage Project's heroic efforts to clear up mysteries and misunderstandings about waste have had so little impact in terms of changing the world of trash. We still waste colossal amounts of food—the EPA pegs food waste in landfills as more than 14 percent of total landfill contents by weight. This isn't much different from what Rathje found more than a decade ago, despite recent attempts to ramp up composting nationwide. Food waste aside, most recyclable materials are not, in fact, recycled. It's frustrating, Rathje said. The problem as he saw it is in how people define the very concept of waste, a question that he said was really more philosophical than scientific.

In modern garbage parlance, Rathje explained, "waste" has become synonymous with "trash"—that is, waste has come to mean the perceived dirty, icky, unhealthful, useless, valueless material that's left over when we're done with something. By this definition, waste is the foul stuff we wish would just disappear. Our entire elaborate waste collection, transportation and disposal system has for a century been built around this "just make it go away" concept, an illusion for which Americans happily (or at least regularly) pay either through taxes or monthly bills. Waste in this sort of discussion is always defined as a cost, a negative and a burden—an inevitable, unpleasant fact of life, for which the only remedy is removal.

But what happens if a different definition of the word "waste" is emphasized—the original verb form of the word, as in "to waste" something? Now the nature of the debate changes, because "to waste" implies the object being wasted has value, be it time, resources or manpower. After all, you can't waste something devoid of value. If trash is defined not as waste but as the physical manifestation of wastefulness, the discussion stops being about disposing of the dirty or useless, and starts being about asking why we are throwing away so much hard-earned money. Why are we wasting stuff that we pay for as product or packaging, then pay for again as trash to be hauled away? Now it's no longer the waste itself that's negative, but the act of creating it that's at issue. And the convenience of burying these discarded items in landfills forever, or shipping them off to China to be recycled for pennies on the dollar (or far less), stops seeming so normal, so sensible.

Rathje used an archaeological analogy to express this distinction between waste and wastefulness. Boiled down to the most simple, broad categories, every great civilization goes through three main stages of evolution. First comes the pre-classic era, the Florescent Period, when a set of small, scrappy villages coalesces into something more powerful, a dramatically rising civilization that has learned how to make a living, be it through warfare, trade, irrigation or some other method of consolidating and capitalizing on resources. Then, having reached a pinnacle of development, the civilization enters its Classical Period, in which it enjoys prosperity, steady growth and dominance. The Classic Maya culture that Rathje studied early in his career featured enormous temples and palaces sprawling across acres of verdant land—classical displays that required enormous resources and manpower to erect. A culture at that stage can afford extravagance. It can be—or at least believes itself to be—unharmed by waste.

Eventually, either through competition from other cultures or simple exhaustion of available resources, a civilization—any civilization—enters an inevitable decline. This is the post-classic or Decadent Period. In ancient Maya, the temples of the decadent years became small, the palaces shrunk, the once treasure-laden tombs grew spartan and poorly constructed. Cultures entering this

terminal phase begin husbanding resources, recycling and repurposing like mad. This is the moment when conservation becomes
the watchword.

But the word always comes too late. Cultures replace extravagance with frugality only after the resources have dried up. Think
Easter Island, the fall of Rome, and any number of empires, from
Persian to Ottoman to Spanish to British. Always, the fall approaches
and the wising-up comes too late.

One of the questions the Garbage Project sought to answer as it
peered in the landfill mirror arose from that tragic history. What
stage, Rathje asked, was American civilization in?

Back in 2001, when Rathje penned an article on this subject for
the surprisingly readable *MSW Management: The Journal for Municipal Solid Waste Professionals*, the answer seemed obvious. The
conspicuous consumption, the outrageous levels of waste, the paltry
recycling rates, the popularity of sport-utility vehicles, the morbid
obesity, the addiction to overpriced bottled water marked up thousands of times over its chemically identical tap water equivalent—
all suggested an America in the midst of a most profligate Classical
Period, embracing the culture of abundance, the illusion of the bot-
tomless well. The headline on his column was "Decadence Now!" In
it Rathje urged what seemed at the time to be a premature embrace
of the values of a decadent culture. America should break the historical pattern and commit to all-out conservation and husbanding
of resources before, rather than after, it was too late. Time to swap
those definitions of waste and wastefulness, Rathje suggested, and
hard as it might be, start thinking about what happens during a
product's end life before we even buy the damn thing. The heedless
wastefulness that has been an American hallmark since the birth
of the disposable economy has to come to an end, he argued. That

would require an act of will, not unlike the decision by alcoholics or addicts to resist their insatiable cravings. "That doesn't come easy, but that's what it takes," Rathje declared. "Decadence now!"

There's just one problem, he added: No great civilization of the past has ever pulled this off. None.

"Can we make a conscious, unprecedented decision to embrace the frugality—the source reduction, reuse and recycling—of the Decadent Period before it's too late, while we're still riding high in the Classic Period?" Rathje wrote. "Will we thereby extend our golden days?"

He wrote that column eight months before the 9/11 attacks. In the decade that followed, judging by the recession-induced shrinking of trash loads heading to landfills, and the burgeoning interest in sustainability, recycling and zero-waste strategies in communities and businesses across the country, it seemed clear to Rathje that we are right on the cusp of our own Decadent Period. Perhaps we've already slipped over into it, he mused, or perhaps we'll pull back. But that drop-off is coming up sooner or later, Rathje predicted, and probably sooner than anyone is quite ready to believe.

"Decadence now!" he said at the close of the interview, then added darkly, "Now or never."

ALTHOUGH RATHJE's Garbage Project ended with his retirement, with no one in the university research world interested in assuming his place as archaeologist of trash, his garbology legacy nevertheless continues. And it is doing so with a decidedly more hopeful spin.

The renaissance comes in the person of Sheli Smith, one of the first students to take part in the Garbage Project—a Moldy Oldy, as the veteran alumni of trash call themselves. It had been Smith who

stumped a game show panel that couldn't guess she was a garbologist, who silk-screened the project members' first official T-shirts
(emblazoned with the image of a hand reaching inside a garbage
can), who braved the derision back when Rathje's colleagues considered him crazed and embarrassing, and when they all referred
to the project as *Le Projet du Garbage*. Even picking through trash
sounds more dignified in French, she says.

After graduating from the University of Arizona in 1976, Smith
went on to specialize in underwater archaeology. This took her as
far from the desert trash sorting scene in Tucson as can be imagined, as she plumbed sunken cityscapes in the Mediterranean and
shipwrecks in the Caribbean. But her work at the Columbus-based
Past Foundation finally brought her full circle three decades later,
when the head of the local Solid Waste Authority had sought the
help of foundation anthropologists. He wanted to design an educational program that could help kids understand and rethink the way
society creates waste. He had no idea he had stumbled on a founding member of the Garbage Project—he had never even heard of it
when he asked if anyone there knew something about waste. Smith
had given him a big grin and said, "Funny you should ask . . ."

Smith led the ensuing effort to create a school syllabus for
an interdisciplinary garbology class project. It started as a public
school pilot with one hundred high school students. They studied
their own trash, their cafeteria food waste, the history of garbage,
and wound up the class with an insider's tour of the local landfill.
The students ended up fascinated and engaged by the hands-on
excursion into a world of trash they never really considered
before—it had been "in sight, out of mind," as Rathje liked to say.
The students were also horrified by this world, as when they calculated that their little school cafeteria wasted sixty-five pounds of

perfectly edible food every day. Then they calculated it would take twenty household composters to handle that load.

"They were stunned. It changed their behavior," Smith says. "They stopped wasting so much food. They demanded the school stop wasting so much."

Based on this success, the garbology program was expanded, reaching first the entire school district, then much of the state's schools. Now it's gone viral. The curriculum, available as a free download, is being picked up for use in classrooms all over the country—adopted, modified, localized. The thing about garbology at that level, Smith says, is that it lets anyone—kids, teachers, parents—understand their own footprint, as well as their friends'. And once that's understood, it's possible to do something about it. Garbology makes it possible for a student to go beyond thinking about saving the world, and actually doing it, Smith says. It's within their power to make a difference.

High school students took it on themselves to renegotiate recycling deals, bringing in more money for their school after they studied their trash flow and calculated the value of their cans, paper and bottles. Third-graders voted to impose a twenty-minute rule of silence at mealtime—because if they concentrated on eating instead of talking, there would be less waste.

"Third-graders did that—it was *their* idea!" Smith says with wonder. "If I had suggested that, they'd think I was some crazy old lady. This is what Bill Rathje made possible. This started with him, and it's still making a difference. It gives you hope for the future."

PART

3

THE WAY BACK

If it can't be reduced, reused, repaired, rebuilt, refurbished, refinished, resold, recycled or composted, then it should be restricted, redesigned or removed from production.

—BERKELEY ECOLOGY CENTER

What the hell was I thinking?

—BEA JOHNSON,
on her pre–zero waste lifestyle

The people who are crazy enough to think they can change the world are the ones who do.

—STEVE JOBS

9 PICK OF THE LITTER

Niki Ulehla laced up her steel-toed boots, pulled on her Day-Glo yellow vest, donned her hard hat and thrust her long fingers with the short, unadorned nails into heavy-duty work gloves—the non-optional fashion statement of the San Francisco Waste Transfer and Recycling Station, aka The Dump. Properly armored, she could begin her day.

She ventures out from her workstation and into the chaotic heart of the dump, an open, drive-in-and-drop-off area known as the PDA—the Public Disposal Area. There she begins searching the piles of twisted metal, scrap wood, broken pictures, crumpled boxes, cracked knickknacks and discarded papers that people had hauled to the PDA from their basements, attics and garages. Something was hidden there, something she needed, lurking within this residue of modern life.

She flipped open a box here, turned over a legless chair there, scanning with a practiced eye the treasures untreasured by luck or death or poverty or time or boredom or age. All of these objects had stories to tell, or so she imagined, but not just any story would do. Ulehla was looking for something particular and unique in those mounds. She was looking for Dante's *Inferno*.

And in fairly short order, she found what she needed, or a piece of it, at least. It came in the form of a stump—a twisted, smallish knob of leathery wood, more shrub than tree, pulled from someone's yard and carted to the dump along with a pile of graying weeds, dried and thorny whips of bougainvillea and crackly brown palm fronds. The stump stood out for her, despite its gnarled condition. It looked like a bird of sorts, perched amid the old papers, the yard cuttings and the mélange of junk. It had a bulbous, beaky head-like projection with a small knot set in the middle like a squinting eye staring straight at her. Winking at her, really. "Okay," she murmured to herself. "I need a bird." She hefted the stump in her gloved hand and considered the possibilities. It didn't look like just any bird, she saw. Its cracked visage hinted at the dreaded Harpy herself—half woman, half avian, the tormentor of lost souls in Dante's second ring of Hell.

Ulehla tossed the Harpy stump into a shopping cart and moved on. There were other things to find, objects to sculpt and paint and bring to life in this loud, odorous place, filled with the roar of diesel engines and the grating beep-beep-beep of big trucks backing up with more loads of trash, more material, more stories. This place was Niki Ulehla's supply house, her crafts store, her inspiration and her muse, for such is the life of a garbage dump artist-in-residence.

She wanted to use trash to create a cast of marionettes to reenact the *Inferno*. This is how she proposed to show the world that exquisitely sculpted works of art could not only be brought to life, but could be crafted from materials that had been abandoned as worthless, unworthy waste. Trash would have a use in her vision, a greatness even, waiting to be tapped.

Tapping that source by redefining waste is the purpose of San Francisco's garbage dump artist-in-residence program. That concept's allure persuaded Ulehla to put her jewelry design business on hold to compete against dozens of other artists for a coveted four-month residency at the dump, where the full-time artistic mission is to expose and exploit the endless uses and potential for the stuff we call garbage. This is the most visible aspect of San Francisco's campaign to put an end to waste and become America's most sustainable and least trashy city. If the *Inferno* could rise from the city dump, Ulehla figured, anything was possible.

She reached into her cart and fingered the stump, imagining how she would carve and paint it, then bring it to life. Then she sighed and renewed her search for the rest of the cast. She had a mere four months to create her art and plan her show, or there would be, quite literally, hell to pay.

THE ARTIST-IN-RESIDENCE program at the San Francisco dump—insiders use the acronym AIR—started back in 1990 as a Southside San Francisco oddity planted a few miles from the airport near the old Cow Palace arena. It has evolved into an unlikely San Francisco icon, frequently copied but outlasting all imitators. Art critics and Bay Area glitterati frequent the AIR shows and receptions, while schoolkids tour "the dump with the art studios" almost daily. A hundred or more applications for the coveted residencies are always on file. There are two artists at the dump at all times, there for four-month stints, for a total of six a year. They are drawn not just from the ranks of promising no-names and up-and-comers, but also established, successful artists eager to make their mark by facing the creative challenge of painting, sculpting, carving, collaging, weaving, welding, writing, photographing, dramatizing and filming trash. Composer Nathaniel Stookey's work *Junkestra* was performed by the San Francisco Symphony with instruments he constructed out of trash. Andrew Junge built a super-realistic life-sized Hummer out of the dump's endless stream of discarded plastic foam; the sculpted car went on to tour the country. San Francisco's central Recycle Center on the waterfront, meanwhile, made an architectural centerpiece out of artist Hector Dio Mendoza's towering seventeen-foot-tall pine tree constructed entirely out of junk mail—a trunk made of drug ads, branches built from credit card offers, and leaves consisting of shredded catalog pages.

The only artistic discipline not yet represented in the dump oeuvre is dance, though not for lack of trying. Every year for the past decade a Bay Area dancer-choreographer has proposed staging a modern dance extravaganza at the dump's most grotesque and busy location, The Pit, which is a warehouse-like structure behind the PDA dominated by a twenty-foot-deep, football-field-sized indoor swimming pool of garbage. An entire day of San Francisco's nonrecycled, uncompostable trash is piled there, crushed by bulldozers, then shoveled into enormous trucks to be hauled to a remote landfill, clearing the way for the next day's incoming refuse tide. It is a loud, stinking, dangerous place, filled with an ever-shifting whirlpool of debris and heavy machinery twenty-four hours a day, seven days a week. No one has figured out how to bring dancers and an audience into the facility with any reasonable measure of comfort or safety.

"Not yet, anyway," says the art program's director, Deborah Munk. Her job description requires her to be open to even the most outlandish artistic possibilities, which she embraces with abandon, as evidenced by the gown woven out of multicolored newspaper delivery bags she wore to the premiere of *Junkestra*. "The Pit is closed two days a year, Christmas and New Year's," she muses. "Maybe we can work something out someday . . ."

The artist-in-residence program was the brainchild of Bay Area activist, artist and environmentalist Jo Hanson. She had bought and renovated an old Victorian in San Francisco's Lower Haight District in the early 1970s, then became a local hero when her one-woman crusade to sweep the trash-strewn street outside her new home grew into a citywide anti-litter campaign. She started to incorporate trash into her art and installations, specializing in using "street-crushed metal" as raw material for her sculptures. Then she started organizing teach-ins and bus tours of illegal dumping sites, hoping to persuade San Franciscans to get greener. In 1990, Hanson visited the Sanitary Fill Company's waste-transfer facility out near the Cow Palace to see where all the litter she collected ended up, and was fascinated by the variety and richness of the materials being thrown away. She felt like digging around then and there for raw materials for her art. Instead, Hanson suggested that the trash company consider sponsoring artists at the dump with stipends, studio space and pick of the litter. They could simultaneously advance the arts and educate the public about waste, while also garnering some positive publicity. If her street-sweeping campaign had taught her nothing else, it was that the news media cannot resist a quirky story about garbage.

Hanson's timing couldn't have been better. A year earlier, California had adopted landmark legislation requiring local governments throughout the state to divert 50 percent of their waste from landfills. Few communities were anywhere close to that goal. With the ambitious trash-to-energy plans of the eighties dead, recycling was embraced as the waste solution of the future. San Francisco, like most other cities in the state, was just beginning to respond to this mandate by ramping up curbside recycling. City officials and the waste companies they hired were desperate to get public buy-in for the concept, and for the long-unpopular chore of separating their recyclables from regular trash. A splashy, high-profile resident-artist program sounded like a great way to further the cause in a town that took pride in both its environmentalist heritage (John Muir, Sierra Club, David Brower) and its support for the avant-garde. Promoting trash art hit all the sweet spots.

More than one hundred artists later, the Sanitary Fill Company, since rebranded with a more eco-friendly name, Recology, continues to support the program with a level of enthusiasm rare in the world of corporate waste management. Director Munk thinks this may have something to do with Recology's own unusual backstory, which dates back to the chaotic independent scavenger guilds that ruled the Bay Area trash business a century ago and that swarmed across the trashed landscape of the city after the massive San Francisco earthquake of 1906. These scrappy, battling trash haulers personified early on the ethic of reuse, repurpose and recycle that are now the rallying cries of modern urban environmentalism—not out of a desire to be green but because in those pre-plastic, pre-disposable-economy years, there was good money to be made from the wood, leather and metal scavenged from the garbage. Emphasizing that part of trash history in its company slogans and reports gives Recology the opportunity to stake a claim as one of the nation's recycling pioneers. It's an accurate claim, to be sure, though the company history omits some of the more unsavory practices of those same early scavengers, such as their penchant for wretched, open-air dumps and trash burning. These many small scavenger companies eventually consolidated into two main city contractors by the 1920s: the Scavenger Protective Association and the Sunset Scavenger Company, both of them employee-owned trash and landfill operations that split the city's garbage, one taking most of the residential areas, the other focused on the business and financial districts. The two city licenses for scavenging granted back in those days are still in effect, though the companies went through several incarnations, name changes, a merger into one corporation called Norcal Waste Systems, a variety of scandals, near bankruptcy and a bribery indictment. Finally Norcal reinvented itself in the new century as a champion of green practices, recycling and the quest for zero waste. In 2009, it completed the transformation by renaming the company Recology (a blending of "recycling" and "ecology").

As of 2011, Recology had contracts for resource recovery and waste services for fifty communities in California, Oregon and Nevada, though San Francisco remained its biggest turf and headquarters home. With 2,100 employees and $351 million in annual revenues, it is one of the ten largest employee-owned companies in the country, and the largest by far in the U.S. waste industry. It's also the biggest organic composter, turning yard waste and garbage from San Francisco's five thousand restaurants (220,000 soggy tons a year) into 150,000 cubic yards of compost that's widely used by the vineyards of Napa and Sonoma Valleys. For San Francisco, that means twenty lumbering trucks that used to haul the stinking, rotting garbage to the landfill are taking it to the composter instead, ultimately returning the food that farmers grow back to farmers in another form—a classic closed loop that brings the natural process of decay back to the human world in a way that landfills never can.

San Francisco boosted these efforts in 2009 by becoming the first major city to collect household food waste at the curb in separate bins along with green waste for composting. Recology was assigned the task to carry out this mandate with specially designed two-compartment trash trucks to keep the organic waste separate on board. It's one of the main reasons San Francisco was able to claim the mantle of green waste leader of American cities in 2010.

City ordinances make the composting mandatory, with violations punishable by whopping fines of $1,000 for each misplaced pile of potato peels and watermelon rinds. Critics of Bay Area progressive politics raised the specter of trash cops snooping through household trash. But so far the toughest response to violations of San Francisco's new garbage etiquette has been by Recology garbagemen, who leave behind a note on offenders' trash cans with a reminder about properly separating waste into the correct blue (recyclables), green (organics) and black (rubbish) bins.

The company did, however, get a court injunction against a small army of independent recyclers who were sneaking at night into neighborhoods ahead of Recology's trucks and pilfering the most high-value recyclables in the bins. This was hurting Recology's ability to make recycling pay for itself, and driving down the percentage of trash San Francisco could claim it was diverting from landfills, despoiling its green credentials.

Recycling may seem like small change, but it's a huge part of

Recology's business model. Recycling theft became big business, too, especially during the economic downturn, with California unemployment hovering at 12 percent since 2009. In an affluent city like San Francisco, poaching recyclables was netting the robbers—and costing Recology—an estimated $2 million to $5 million annually. The injunction authorized police to make arrests and imposed penalties of $1,000 and six months in jail for each instance of poaching; the organized gangs of trash thieves soon moved on to easier hunting grounds. (Reflecting the national scope of the trash poaching, a number of other cities, from Sacramento to New York, have adopted comparable measures against recycling thieves.)

With the advent of the injunction and the food-waste recycling, San Francisco claimed in 2010 to divert a nation-leading 77 percent of trash away from landfills through Recology's recycling and composting operations.

Still, the city fills up The Pit with 1,500 tons of trash every day from residential "black bins"—the stuff that's not recycled. Periodic trash audits show that two-thirds of the material in The Pit is a mix of plastics and food waste, which means it could theoretically (though not economically) be recycled, too. Instead, it gets hauled by sixty huge diesel eighteen-wheelers a day to a Waste Management, Inc., landfill in Altamont, fifty-eight miles away. The impact of these waste shipments cancels out many of the environmental gains of San Francisco's efforts to lower the city's trash footprint.

In 2015, the city plans to shift gears and haul that waste to a Recology landfill near Sacramento by rail, lowering transportation-related emissions considerably. By 2020, the city's master trash plan calls for zero waste to landfills, though it's not entirely clear yet exactly what that will look like, or whether it's possible to achieve.

And despite these impressive efforts, an underlying reality is

APPLICATION FORM

Applications are due September 2, 2011

Mailing Address
AIR Program
Recology San Francisco
501 Tunnel Ave. San Francisco, CA 94134

First Name   Last

Address

City   State   Zip

Phone

E-mail

Website URL

Professional Activities
Indicate if you teach art, work as a curator or have another art-related job.

Tools
If you are selected, what tools would you use?

- ☐ MG Welder
- ☐ Oxyacetylene system
- ☐ Plasma cutter
- ☐ Horizontal band saw
- ☐ Shear, brake, roller
- ☐ Glass kiln
- ☐ Tile saw (wet saw)
- ☐ Sewing machine
- ☐ Small etching press
- ☐ Darkroom

- ☐ Table saw
- ☐ Wood lathe
- ☐ Scroll saw
- ☐ Router table
- ☐ Drill press
- ☐ Bench sander
- ☐ Band saw
- ☐ Miter saw
- ☐ Computer
- ☐ Video camera

Other
Additional tools or equipment you would bring.

Misc.
When did you take a tour of Recology San Francisco?

From the application for trash artist-in-residence, Recology, San Francisco:

Program Goals

- To encourage the reuse of materials
- To support Bay Area artists by providing access to the wealth of materials available at the public dump
- To prompt children and adults to think about their own consumption practices
- To teach the public how to recycle and compost in San Francisco through classroom lessons that explain the city's three-bin (recycling, composting, garbage) system

Recology Provides

- Twenty-four-hour access to the facility
- A large, well-equipped art studio
- An exhibition and reception at the end of the residency (including printed invitations, refreshments, installation assistance, etc.)
- Miscellaneous supplies and equipment
- A monthly stipend
- The presentation of artists' work in off-site exhibitions

Expectations of the Artist

- Work in the studio either forty hours per week for a full-time residency or twenty hours per week for a half-time residency
- Greet and speak to tour groups during weekdays and on the third Saturday of the month
- Make three pieces of art for the company's permanent art collection

- Use materials recovered from San Francisco's waste stream
- Be available to talk with the media
- Leave all art created during the residency with the company for the next twelve months for exhibitions at off-site venues

that keeping trash out of landfills is not the same as making less trash in the first place. Indeed, San Francisco residents tend to make slightly more waste per person than the national average. So if anything, the knowledge that most trash is being recycled or composted may be giving San Franciscans license to be more wasteful rather than less.

"In any case," says Munk, "there's still plenty of room to improve."

FINE-TUNING THE recycling process, knocking down the flow of stuff into facilities like The Pit, keeping the poachers in check— that's all good and sensible, Munk says. But she figures the next strides, the closing of that last big gap between taking less to the landfill and making less waste, will require something beyond changes in industrial trash collection methods and destinations. It calls for changes in mind-set, in how regular people think about what they regularly buy—or don't buy—which governs what they do and don't throw away. And that's the essence of Munk's job: Changing minds, perceptions and comfort zones, she says, is what the artist-in-residence program is supposed to be about.

Munk ought to know—it changed hers. She was a clothing buyer for a high-end San Francisco boutique. It can be a lucrative living, making your work and your personal life revolve around consumption. But she eventually decided there was only so much stuff you can cram into a closet, or a life, before it gets old. Surprising her friends, family and herself, she gave it up in a heartbeat when she bumped into one of her old college professors on the street in 2000, and he mentioned that he had just taken what he considered to be the coolest job he had ever heard of: running an art program at The Dump. Munk had never heard of the residence program before, but almost without thinking, she blurted that she needed something new to tackle, too, and she wanted to return to the arts and education studies she had pursued in college. Could she come work for him?

"You can start next week," he replied. She did, joining the art staff at The Dump. She worked part-time at first, still supporting herself as a fashion buyer, her life an uneasy truce between opposing values. Within a year she came on full-time and eventually succeeded her professor as director in 2007.

Her shared office is filled with posters, photos of trash art, a few oddities from The Dump and the most vital objects in the room, a mass of black three-ring binders. Each represents a single artist's application for a residency, shelves of them dating back to 1990, containing proposals for their projects, personal statements, résumés, work samples, portfolios, recommendations, pleas. There's also a checklist of required artists' tools only a dump denizen could love, ranging from sewing machine to band saw to metal inert gas welder and plasma cutter. The binders are thick and detailed, and document a surprisingly intense passion to tackle trash that can't even begin to be explained by the $1,800 monthly stipend. Munk and her two colleagues make the first cut of applicants, then a community advisory board makes the final decisions.

"I know what to expect when I heat a piece of steel ordered from a supplier," wrote the first artist-in-residence, sculptor William Wareham, whose job included setting up the first rough-hewn studio at The Dump in 1990. His rough-edged sculptures of torn metal and old shopping carts proudly displayed their trash provenance. "But with this, it's impossible to know with certainty. That's one of the things that makes this an exciting project."

A year later, Remi Rubel's large and colorful quilt-like mosaics of bottle caps and other found objects looked like anything but trash. Yet she, too, seemed to be a natural for the program: "From the day I learned to walk," she wrote, "I began searching the ground for treasure. Tumbled glass from the sands of Lake Michigan or coins from sidewalks later became bottle caps and flattened cans from the streets." The Dump became her ultimate treasure hunt.

Susan Leibovitz Steinman designed a hilltop sculpture garden behind The Dump in 1992, where each artist—they were all sculptors at first—was asked to deposit at least one work. (The landscaped hilltop garden also served a practical purpose: a noise buffer between the clang and crackle of the transfer station and the Little Hollywood neighborhood next door.) Winding through the garden, Steinman placed a path of crushed gray concrete recovered from the Embarcadero Freeway that collapsed in the massive 1989 earthquake; it's embedded with found objects and the words of students contemplating the future. She called it the River of Hope and Dreams. "It feels really good to me to rescue things. Things come with stories already in them," Steinman said.

Towering over one part of the garden is Marta Thomas's striking "Earth Tear," a curving, eight-foot-tall teardrop made of individual plastic bottles that seem to flow like liquid across their rust-flecked rebar frame. All scavenged, the common pieces of refuse have been shaped into something magical, symbolizing for Thomas sadness at environmental harm assuaged by the hope for renewal that grief can usher in. And like the plastics adrift in the ocean, Earth Tear's plastic bottles have begun to break down, clouded, cracked and brittle from exposure to sun, mist and rain.

In its second decade, a broader range of the arts was represented in the mix of residents—painters, videographers, graphic artists, musicians. The Dump was nothing less than a playground in 2010 for artist Ben Burke, founder of San Francisco's Stars and Garter Theater Company and Apocalypse Puppet Theater, who put his thoughts about the program to verse:

> It rains and it shines, the world turns on a dime, and our grease is the trail that we leave. We spin yarns to the moon, for the story's a loom, it's the carpet we walk that we weave. So go on, act the fool, for the sea is not cruel, and the ship, it turns out, is not sunk. It's just run aground, as the table spins round, and it's time to build fables from junk.

THE TWO artists of summer 2011 provide a fascinating contrast as they work in adjacent sections of their warehouse studio. The raucous sounds of the transfer station provided background music for their labors, seasoned by the occasional shout of an air horn from a passing train on the freight line across the street. The ever-changing trash storehouse of the PDA is just steps away, a tide that washes new material their way throughout the day. From their studio, they can see something interesting as it arrives, and leap into action.

Of the two artists, Abel Rodriguez is a bundle of energy, constantly in motion. He doesn't keep a chair in his crowded workspace. He likes to stand as he works, moving among the hundreds of items he has rescued from the piles—the finely turned legs from a cracked and broken table and chair set, the pieces of a wooden dish dryer, a pack of blank CDs still wrapped, a brand-new tea ball, silvery bent hubcaps, a photograph of Emiliano Zapata fresh from some revolutionary battle surrounded by his troops, who look all of twelve years old, not much bigger than their rifles. Rodriguez says he already has accumulated too much material, an abundance of artistic riches, but he cannot help going out every day to see what the trash tides wash his way. When he agrees with something you say, he answers in rapid fire: not "Sure!" but "Sure, sure, sure!" He's a fast talker, a streamer, and his thoughts come in a rush. "Nothing's ever still here, nothing's ever fixed, everything's in constant motion. That's how it's been all my life, so this environment is perfect. Perfect. I don't put things away, I don't like to waste. I like to repurpose. Waste is a misnomer."

The Yale University–trained artist certainly incorporates this ethic in his work, big collages and sculptures that resemble the sorts of amalgams of debris that wash ashore, fragile and inviting. They have a deliberately temporary quality. He likes to attach things with visible applications of tape to emphasize impermanence— as well as to let him pull things apart and reconfigure them on the fly. "Nothing is permanent," he says, pacing his part of the studio. "Nothing, nothing, nothing."

Lauren DiCioccio is the perfect complement to Rodriguez, her energy contained versus his constant motion. "I sit," she says, laughing. "I pretty much have to."

DiCioccio chose the extremely time-consuming fine work of embroidery, sewing and textiles for her residency. Her art depicts in fabric the common, and sometimes uncommon, objects she pulls from the trash heap—photo albums, cookbooks, a torn baseball, a dead mouse (okay, she concedes, she didn't actually bring the real version of that back to her studio), a box of Kodacolor film, a post-card sent airmail. They are life-sized and compellingly realistic, yet whimsical cloth versions of reality. She particularly likes paper objects—newspaper pages, letters and maps, for which she pains-takingly sews printing, handwriting, even the blue lines of ruled loose-leaf paper. These familiar objects, she says, are "obsolescing" right before our eyes, being replaced by digital and virtual alterna-tives, or simply falling from favor. Capturing them with embroidery imbues them with a kind of poignancy. She usually leaves long threads dangling from her sewn objects, as if their place in the world, their very reality, is unwinding and coming apart before our eyes. People want to touch DiCioccio's art, which is an art-lover's no-no, of course. But there's something so personal, so playful yet bittersweet, about her work that visitors just can't help themselves. She's good-natured about it, saying that those urges show that she has succeeded in her work, and tries to channel the enthusiasm by putting several pieces in a touchable display—with gloves next to them for art fans to put on first, to avoid dirtying the objects.

"I have had a total crush on this program ever since I heard about it," DiCioccio says. "I had to come here. I loved the idea of being able to see a portrait of the lives of the city of San Francisco in the things that people throw away. I'm drawn to nostalgia, even when it's other people's."

She was especially drawn to one object, a pair of hand-tinted photos of a man and a woman. They're probably husband and wife, though that's just a guess—there were no captions or notations on the back. The pictures were joined in a double-paned leather frame hinged like a book. The hand-tinting, a technique common in the fifties and sixties, now a nearly lost art, was beautifully done, and the faces seemed to glow as if illuminated. They were smiling, he more broadly than she. Clearly, the photo display had been a beloved keepsake, until one day, it was not, and the trash heap claimed it. "I can't stop looking at it. It has a homely quality, and yet they're just so beautiful. Clearly this was somebody's treasure. I can't stop thinking, How did it end up here?" DiCioccio's rendition with needle and thread somehow captures the sadness and mystery of that transition, its threads hanging.

"One of the interesting things to me is when you go out there and you find these things, and you realize this person has probably died, and this was probably what was in their top bureau drawer," DiCioccio says. "These objects *are* that person in a way, it's their spirit. And I'm rooting through some very personal, bittersweet, touching things, and trying to bring new life to them—to rediscover the human qualities in these disposable things . . . If we can do that, it's not trash anymore, is it?"

WHEN ARTISTS-IN-RESIDENCE first arrive at The Dump, they invariably fret that they'll never find the things they need for their work: that particular piece of wood, metal or plastic, that just-right shade of paint, that certain thread or cloth. But the fears are almost always unfounded. The flow and variety of materials in the trash are always far greater than any of them quite believe—until they actually start wading into the stuff, and they realize the house cliché is uncannily true: *If you need it, it will come.*

One artist needed an array of nuts and bolts to assemble his 3-D collages, and thought for sure he'd have to resort to the hardware store. But then someone showed up at the PDA with the contents of a home workshop, jar after labeled jar of screws, nuts, nails and washers of every size and shape, labeled and lovingly organized, thousands of them. Those baby food, mayonnaise and pickle jars of hardware have served successive artists for years. Someone else needed thread and needles, and really, how likely was it those small essential elements would turn up? How would they ever find those proverbial needles in this stack of garbage and mess? And then there they were, not one but two sewing boxes from different "donors," crammed with old, good thread, old-school wooden spools, grandma's sewing kit transformed into trash, and then transformed again into art and the tools for art. Sooner or later, what the artists need always shows up — cloth, glue, rubber, paint. Paint, rather shockingly, shows up by the hundreds of gallons, so much so that Recology sends truckloads of it overseas to developing countries to paint schools and clinics, as well as offering it free to any locals who want to come down and pick it up. There's still plenty left over for the artists. Who knew so much good paint ends up as trash, oftentimes cans of it barely used or never even opened?

Here the waste becomes another kind of abundance, and the artists become scavengers, archaeologists, miners and collectors — fascinated and horrified by useful things that somehow have been relegated to the trash pile. Recology has an entire side of a huge warehouse reserved for stuff that's too good to throw away — furniture and appliances that are usable, sometimes even gorgeous, that people have hauled to the dump. Recology redirects thousands of such pieces a month to a network of thrift stores — unless the artists get to them first.

The fact that galleries of art, and important milestones in the careers of professional artists, can be constructed out of ordinary trash reveals much about the artistry of the residents, Munk says.

But it reveals even more about our wasteful ways, and how readily we attach the label "trash" to perfectly good and frequently beautiful things. How much of our trash, how much of our 102-ton legacy, is really and truly trash? At San Francisco's artist-in-residence program, they'll tell you: not nearly as much as you think.

THE INFERNO performed by trash puppets was, Deborah Munk recalls, a rousing success. People were fascinated by the drama, which featured neither dialogue nor narration, but just the expressive gestures and interactions of the carved trash-marionettes. Niki Ulehla is more modest, saying only that she needs to work on the show a bit more and would like to try again in a year or two. But she allows that the audience appeared to enjoy the performance.

Her puppets are eerily, absurdly lifelike. A twitch of the controller—that x-shaped wooden cross that lets a puppeteer direct the limbs of a puppet—and Ulehla can make Harpy walk. Not bounce along in cartoon simulation of walking, but truly walk, an evil, ominous shuffle appropriate to a tormentor from Hell. She apprenticed in the Czech Republic, where puppetry and marionettes are still a vibrant art form, and for ten years she has been honing the technique she learned there. Her puppetry gives her a special credibility with the five thousand schoolkids who troop through The Dump every year to see where the trash goes—and what the artists can do with it. This, Ulehla says, was her favorite part of the artist-in-residence experience.

"I guess the small part that I was able to play here was to show by example that there's an alternative to putting something in the trash. There's an alternative to buying more materials. We can do better with our resources. We can bring new life to what we thought was disposable."

As she is telling me this, in an endearingly shy and reticent way, Niki Ulehla is playing with Harpy, making her creation shuffle and turn and peer up at me. With one eye wide and the other slitted in a rakish wink, the malevolent puppet does not share its creator's shyness. "When you spend time here, it's so obvious. So much of the trash really isn't trash. It's staring us in the face."

10 | CHICO AND THE MAN

NOT TOO MANY PEOPLE WHO HEAD TO THE LAND-
fill with a pile of weeds, deadwood and yard waste end up coming
home with a new business, a new mission in life and a new super-
villain alter ego. But that's Andy Keller and ChicoBag's story, one of
those apocryphal origin tales about a regular guy's green awaken-
ing that just happens to be true—and that also thrust him into the
middle of a war between bag manufacturers and a growing list of
cities that want to ban the plastic bag for good.

Keller used to be in the software trade, selling enterprise-grade
code for a company in San Francisco. His employer let him telecom-

mute from his home in Chico, California, in the Sierra foothills north of Sacramento, where you can get a lot of house, even if it is a fixer-upper, for the price of half a condo in San Francisco. But in 2004, his company was bought out by a European conglomerate that wanted him back working in the mother ship. It was too far to drive every day, so Keller had a choice to make. Should he stay with a company that would likely be downsizing once it finished cracking down on telecommuting? Or stay in Chico and find something else to do?

At age thirty-one, single, with no family to put at risk by being out of work, he decided to take a severance package instead of moving. Then he finally started the long-postponed fixing of the fixer-upper, beginning with the landscaping. He had to figure out what he'd do next, and thinking about it while digging in the dirt outdoors seemed a good choice after years of pounding keyboards. At the end of the day, he had a huge load of yard waste with no place to put it, which led to his first-ever trip to the local landfill.

The first thing about the dump that hit him was the odor—nasty, he thought, but unsurprising. Then there was the sight of a mound of the day's trash piled high. Impressive, awful even, but again, no surprise. And then he saw the plastic bags, flashes of white blowing around the landfill, catching on tractors, on gears, on fences. Birds were pecking at them. They seemed to be everywhere, in and around the trash pile, the most identifiable item in the landfill. That, for some reason, did surprise him. Prior to that moment, he had not thought of those handy-dandy filmy white grocery bags as any sort of problem. They were so thin, so light, he hadn't really given them a thought. But their footprint seemed magnified now by their dramatic presence in the landfill.

It struck him so hard that, by the time he was driving away from the landfill, he was thinking that he ought to start avoiding those bags in the future, maybe start using those cloth reusable bags some shoppers brought to the market. On the ride home, the trash by the side of the road he had barely noticed before now seemed much more visible. He saw quite a few plastic bags in the mix of litter.

By the time he got home, he didn't just want to cut plastic bags out of his life. He was thinking this impulse could translate into his new job—creating a good alternative to disposable plastic bags. Of course, there were alternatives already being used by a minority of shoppers, even in those days before communities had curtailed, taxed or banned plastic shopping bags. The problem was, Keller had never really liked those reusable shopping bags. They were bulky, they didn't fold up well, they were so inconvenient that people were forever forgetting them at home or leaving them in the car. They had never really enticed him before, and he guessed others felt the same. A sensible product with great environmental benefits had been stymied by poor design. So what if he designed a bag that *was* convenient? That could fold up so small that you could keep it in your pocket or your purse and never leave it behind again? It hit him like an adrenaline rush, he would later recall: the solution to his unemployment, to his entrepreneurial yearnings and to an environmental and natural resource problem.

He'd make a better bag.

The next day he went to a thrift store and bought an old sewing machine, purchased the lightest, thinnest cloth he could find at a fabric shop, set up shop on his kitchen table and started making prototypes. Then he raced back to the thrift store and bought a better sewing machine—one that actually worked—and spent days sketching, cutting, sewing, then returning to sketch a different design, and starting over.

Keller wasn't your average software salesman (or your average thirty-something guy in any profession) in that he had a number of low-tech, traditional skills his peers tended to lack, sewing chief among them. This was a direct result of his parents declining to give him an allowance. By age twelve he was a full-fledged entrepreneur, always scrambling to make money—cutting lawns for a dozen clients, taking on handyman work in the neighborhood, painting, cleaning yards and selling homemade Santa hats he figured out how to make on his mom's sewing machine.

The shopping bag prototype he settled on was rough and inexpertly sewn, but it was a proof of concept, a template to begin manufacturing for real: a bag that folded up into a pocket-sized sack that closed with a drawstring. He named it, and the company he formed, ChicoBag, patented his invention and financed the launch with an $80,000 line of credit on his house and a credit card that soon maxed out at $5,000.

The business grew slowly but steadily over the next several years, with the original ChicoBag, made of woven polyethylene plastic, selling for $6 retail. Once the business was established (and competitors in the $100 million reusable bag business came out with their own pouched grocery bags), Keller began to expand the company's product line bit by bit, moving into daypacks, duffels, messenger bags and mesh bags for produce, each offering a variety of material choices, including hemp and recycled polyethylene. The original $6 bag, for example, has a greener counterpart made with 97 percent recycled materials for $9.60. The higher price demonstrates the difficult economics of recycling, which is one big reason why there is such a big disparity between what *could* be recycled and what actually *is* recycled.

Manufacturing is done by contractors in Vietnam and China, with some detailing work in the States, which means ChicoBag products have a carbon footprint from transoceanic travel that homegrown products would not carry (though the homegrown versions would carry an untenably larger price tag). Keller says he doesn't try to hide this fact the way some companies do by having final assembly in Guam so a "made in America" label can be slapped on legally.

"That's the reality of manufacturing in America," he says. "Mostly, it's not in America. And even if we did make the bags here, we'd have to import the fabric from China, because that's where it's all made."

ChicoBag's pitch is that a reusable bag, even made from virgin rather than recycled materials and imported from China, is vastly superior ecologically to the disposable alternative. For the average American, that alternative amounted to five hundred or more plastic bags made from fossil fuels thrown in the trash each year.[1]

By 2011, the company had expanded from Keller and his kitchen table to thirty employees and $5 million in annual revenues. ChicoBag had 14 percent annual growth in 2010. Competition is fierce in the reusable bag business—there are twenty rival bag makers in California alone, and they're all wrestling over the mere 5 percent of Americans who choose them over disposable plastics—although that percentage is growing as increasing numbers of communities move to limit or ban disposable grocery sacks. Keller's business plan focused on three revenue streams to keep the company moving: retail sales, custom orders (bags bearing the logos of nonprofits, universities, conferences and businesses) and educational sales.

It's this last part of the business, where salesmanship meets green advocacy, that gets to the heart of Keller's decision to go into business for himself making bags that, when rolled up, look like lumpy little Hacky Sacks. This is also the part of the business that almost put him out of business, making him public enemy number one for the makers of plastic bags.

It started simply enough: Keller approached schools and suggested they rethink their community fundraisers by selling an eco-friendly, useful product—namely ChicoBags emblazoned with a school's name and slogan—instead of selling candy and cookie dough in an age of childhood obesity. Quite a few schools took him up on the offer.

As part of this growing school side of his business, and also to promote the virtues of reusable bags over disposable plastic, Keller began passing on educational information about plastic's environmental footprint, waste and cost. He made school visits and suggested that students bring to class all the plastic bags they could find at home, tie them together, and see how far around the school building the chain stretched—a feat that usually left students and teachers alike stunned by their immense consumption of bags.

Then he did a few simple calculations for them and showed that if every American did that with all the grocery bags they used in a year, the chain would circle the earth. Not once, but 776 times.[2]

Next he started carrying around five hundred single-use bags (grocery bags, produce bags and other disposable sacks) jumbled together in a big ball to represent the average American's yearly plastic bag habit. It was big enough for him to crouch down and hide behind, then jump up and startle passersby. He called this a "Candid Camera moment" that got people to laugh, then ask what he was doing. That was Keller's opening to explain how that ball encapsulated what he saw on his first-ever visit to the landfill—and how it also represented the average American's plastic bag use for one year. The five-hundred-bag figure was a conservative estimate, he'd say. The real number was probably higher. Counting just one-use grocery shopping bags, Americans were collectively consuming 102 billion bags a year.[3] The count goes up to more than 150 billion when all types of plastic bags are included. And since every five hundred bags represent the energy equivalent of a half gallon of gasoline, that meant our disposable plastic bag habit was costing us the same as burning 150 million gallons of gas each year.[4] It was all so wasteful, Keller would say, and yet so avoidable.

"The reaction I would get," Keller recalls, "was almost always: *Oh my God, I had no idea.*"

Eventually, he came up with a super-villain alter ego, Bag Monster. Instead of five hundred bags in a big ball, he made a costume out of them and wore the five hundred bags himself, becoming a roly-poly elastic Medusa festooned with streamers of plastic, his face the only visible human element (often contorted into wild expressions that gave him more than a passing resemblance to actor Jim Carrey). Bag Monster loved plastic bags and urged everyone to become a Bag Monster, too. The effect was disturbing, hilarious and irresistible. Schools loved him, requesting visits and demonstrations. Then green conferences and other venues and events started asking to borrow the costume. Keller made up several of them and started a free Bag Monster lending program after perfecting the costume design. (First-generation Bag Monster consisted of bags sewn to a graduation gown, which got hot and sweaty fast, after which Bag Monster got a bad case of BO. This was because the costume couldn't be washed—the bags would just shred in the laundry. Second-gen Bag Monster consisted of Velcro strips with bags sewn on, which were then stuck on a jumpsuit. The strips of bags could be taken off and set aside, allowing the jumpsuit to be laundered, vastly improving Bag Monster's personal hygiene.)

Demand was so great Keller ended up fashioning a hundred Bag Monster costumes. To go with them, he developed a ream of informational materials on single-use plastic bags, the history of plastic and the positive impact reusable bags could have. He put all this information on the company website as well as a separate BagMonster.com site to promote the end of single-use bags. Then he launched a nationwide Bag Monster tour of twenty cities where officials were considering plastic grocery bag bans. He drove around in a van emblazoned with a life-sized picture of his alter ego and a sign pleading with citizens to: "Help Stop the Bag Monster!" The tour culminated in August 2010 with Keller and ninety-nine friends—each wearing one of the one hundred Bag Monster costumes—descending on San Francisco en masse. They marched from the major Bay Area chocolate and chowder tourist spots of Ghirardelli Square to Fisherman's Wharf and back. Then they held a press conference touting a statewide ban of single-use plastic bags, which the California Senate was debating at the time. The *New York Times* interviewed Keller. The TED (Technology, Entertainment and Design) Conference invited him to speak.

And that's when the trouble started. Bag Monster, it seemed, had gone too far. And the plastic bag makers, whose spending on lobbyists in Sacramento alone dwarfs ChicoBag's annual profits— decided to fight back. Three plastic bag makers filed a lawsuit and launched a public relations campaign intended to reveal Keller's environmental message as nothing more than deceptive advertising and self-serving propaganda. It was time, the makers of plastic bags decided, to muzzle the monster.

HISTORY OF THE PLASTIC BAG

1957: Plastic sandwich bags are introduced to replace wax paper.

1958: Plastic dry-cleaning bags replace brown paper.

1959: After eighty children suffocate by plastic dry-cleaning bags, California tries to ban them. Industry lobbyists succeed in killing the ban in favor of a product warning label.

1961: The Keep America Beautiful antilittering campaign is launched with disposable-product industry funding, placing the blame for trash and pollution on consumer litterbugs rather than on manufacturers.

1966: Plastic produce bags on a roll replace brown paper sacks in grocery stores.

1974–75: Sears, JCPenney, Montgomery Ward and other big retailers replace paper with plastic bags.

1977: Paper or plastic? The plastic grocery bag is introduced to the supermarket industry.

1988: Suffolk County, New York, passes the first ban of plastic grocery bags. A suit by plastic bag industry trade groups overturns the ban.

1990: Maine bans single-use plastic bags in retail stores, but the law is overturned.

1996: Four of five grocery bags are plastic.

1997: The name "Great Pacific Garbage Patch" is coined and brought to the world's attention by Charles Moore and his Algalita Marine Research Foundation. They report that the plastic used in grocery bags is one of the most common found at sea.

2005: San Francisco proposes the nation's first tax on single-use plastic bags—17 cents, the estimated cost to society and tax-payers of dealing with plastic bag waste.

2006: An industry-backed provision of the California Plastic Bag and Litter Reduction Act outlaws the sort of fees proposed in San Francisco.

2007: San Francisco bans single-use plastic bags.

2010: At least twenty communities nationwide follow San Francisco's lead, either banning plastic grocery bags or imposing a fee on their use.

Sources: ChicoBag and the Packaging Institute

THE FAMILIAR plastic grocery sack with the two loops for handles began its conquest of the carry-home, take-out world in the U.S. in the early 1980s, after Mobil Chemical (now ExxonMobil Chemical) sued to overturn a Swiss company's patent on the bag's design. Mobil won the case and the right to make the bag (as did anyone else who cared to do so) and the plasticky floodgates opened, helped by the well-timed invention of a machine that could churn out those same thin, white bags at the startling rate of five hundred a minute. The demise of the patent and the near-simultaneous advent of such speedy mass manufacture ended what had been the venerable paper grocery bag's cost advantage and, therefore, its industry dominance. Weight, price point and ease of shipping (one truckload of plastic bags had the grocery-carrying power of four trucks of paper bags) all were suddenly in plastic's favor. And if plastic needed more of a boost, paper companies in that era already had been taking a beating on the issue of deforestation and endangered species.

Even so, consumers at first wanted to stick with the familiar, sturdy, traditional brown paper sack. Paper bags held more, they stood up on their own, you could line them up in the trunk or the backseat of the car. Try that with plastic bags and everything flops out and rolls around. Customers hated them.

But then, consumers hated pretty much every plastic thing compared to whatever material the plastic was imitating or supplanting—at first. The manufacturers knew what would happen next, though. New habits would set in, old objections would fade away and pretty soon consumers would start behaving as if there had never been anything other than plastic laminate kitchen counters or vinyl chair cushions or car bumpers made of polyolefin or bags made of polyethylene plastic film. All the bag makers had to do was flood the market, exercise a little patience and let the unnatural take its course. That's long been the bottom-line truism of the industry: If you plasticize it, they will come—whether they want to or not. That's how you build a disposable economy, and it had worked since the 1950s.

An overturned patent and speedier plastic bag machines, coupled with the cheap natural gas of the era (plastic grocery bags are made from a methane by-product, not oil), meant the "Paper or plastic?" tide inexorably slid in favor of those flimsy white sleeves during the 1980s. Plastic grocery bags moved from a 4 percent market share at the beginning of the decade to more than 50 percent by the end of the eighties.

This played out amid dueling PR campaigns between the Grocery Sack Council (twenty-six plastic bag companies) and the American Paper Institute, each accusing the other of hawking inferior products and harming the environment. There were some early successes by the paper industry and brief flirtations with plastic bans. In 1990, Suffolk County, New York, banned plastic bag use in grocery stores but not other retailers, a differentiation that had less to do with the environment and more to do with the fact that a major Suffolk County employer happened to manufacture plastic bags for non-grocery retailers. Plastic grocery bag makers sued over the unfairness of this and won, overturning the law in less than a year. Also in 1990, a unique piece of legislation took effect in the state of Maine that stated: "All retailers in Maine shall use paper bags to bag products at the point of retail sale unless the consumer requests a plastic bag." Maine became the one state in the union to change "Paper or plastic?" into "Paper (unless the customer speaks up and insists on plastic)." Maine also just happened to be a state in which the dominant industry and single largest employer at the time was paper and timber, and where the governor's brother was a paid lobbyist for the American Paper Institute. The outcry over this bit of heavy-handed favoritism was so great that the law ended up repealed the following year. Stores could bag however they wished, with the only extra requirement being a mandate for supermarkets to have a recycling bin for customers to stuff their old plastic bags into if they felt like it. After that debacle, the paper-plastic war pretty much went plastic's way for the next fifteen years.

By the start of the twenty-first century, plastic bag manufacturers controlled more than 90 percent of the grocery bag market. The plastic industry is one of the few U.S. manufacturers that has not completely offshored itself, employing a million and a half American workers in recent years. It is able to wield considerable political clout through a web of trade associations and groups, including the powerful, 140-year-old American Chemistry Council, the defender-in-chief of all things plastic as the ultimate in safe, cost-effective packaging.

Certainly grocery bags are among the most common and simplest of plastic creations, most often made from high-density polyethylene, which in turn is made from the fossil-fuel-derived gas ethylene. A relatively low percentage of plastic grocery bags are recycled, about 5 percent nationwide despite being theoretically 100 percent recyclable. That's because recycling the bags is notoriously difficult, as the lightweight filmy sacks clog the machinery and end up being carted to landfills as recycling "residue" more often than not. The bags cannot be recycled endlessly, contrary to common belief, but can only be "down-cycled" one time into some product other than bags. (Paper bags are far from perfect and the question of which does more environmental harm, paper or plastic, is still open to debate—plastic bags are greater litter makers, but paper bags are more energy and water intensive, and therefore have a greater greenhouse gas footprint. On the other hand, paper bags can be endlessly recycled and, unlike recycled plastic, recycled paper bags are cheaper than virgin materials.)

Up until 2005, the recycling figure reported by the EPA for plastic bags of all types was 1 percent. After that year, the EPA only reported the recycling rates for all bags, sacks and wraps, a much larger, mixed category of products, for which the recycling rate in the most recent year reported was 9.4 percent, better but still anemic. The plastic industry, not the EPA, is the primary source for these recycling figures; the effect of the change, intended or not, was to conceal the true recycling rate for bags as a discrete category. And even the combined figure is suspect; the true recycling rate may be significantly lower than 9.4 percent. As the Columbia University/*BioCycle* biannual study of the waste stream has shown, EPA reports consistently underestimate the amount of trash made in America, while overestimating the percentage of trash that gets recycled. So the paltry 9.4 percent recycling rate represents a best-case scenario.

Around the time that Andy Keller started up ChicoBag, concern had begun to mount about the environmental impact of plastic bags in general, and single-use grocery bags in particular. These concerns—about ocean and river pollution, litter, impacts on wildlife, the cost to taxpayers of disposing of plastic bags in landfills, the ineffectiveness of recycling efforts for a product that was supposed to be 100 percent recyclable—led a number of local, state and national governments to take action. Which in turn led the plastic industry and the American Chemistry Council to fight back just as they had done with the paper industry.

Ireland was among the first governments to act, way back in 2002, and that country's success became a model for the rest of the world. A plague of plastic grocery bags papered Ireland's coveted green countryside, the Emerald Isle's reputation for physical beauty marred by roads, gutters, foliage and trees sporting beards of windblown bags. Ireland's lawmakers and voters decided they had had enough: They passed a plastic bag tax of 15 euro cents (raised to 22 euro cents in 2007, the equivalent of 30 cents U.S.). The rationale for the new tax boiled down to this: Plastic bags are a great product, but they have been used and disposed of in a profligate and wasteful way. This is because the bags have been viewed as "free" by consumers, when in fact they cost quite a bit in terms of the burden they impose on the environment and on taxpayers, who have to foot the bill for litter cleanup, landfills and pollution. The tax simply reflected the true costs of the bags—in effect, proponents argued, it merely ended a kind of "plastic welfare" that had acted as a market force that favored waste and unproductive overconsumption. The tax proceeds went to Ireland's environmental agency.

Fifteen cents is a relative pittance, but it does add up over hundreds of bags, and so it had an immediate and profound impact on the behavior of Irish consumers. They simply didn't want to pay for a bag that had previously cost them nothing. Plastic bag use dropped more than 90 percent in a matter of weeks. Reusable bags became fashionable, while carting around plastic bags increasingly was viewed as a social gaffe. People would stump out of the store with loose cans and loaves of bread cradled in their arms and then dump it all in a heap in the trunk rather than buy a plastic bag. The tax was never much of a burden because relatively few people paid it—they simply avoided the bags. The windblown litter was curtailed. The main question among the Irish was why it had taken so long to come to their senses.

Predictably, grocery chains acted as plastic manufacturers' surrogates and opposed the levy. But they soon changed their positions for a variety of reasons, not the least of which was that it turned out to be good for business. "I spent many months arguing against this tax," recalled Feargal Quinn, who founded Ireland's top grocery chain. "But I have become a big, big enthusiast . . . Now we're saving the environment, we're reducing litter and since we're not paying for bags, it ultimately saves money for us, and that reduces the price of food for our customers."[5]

For the first time since the dawn of the age of plastic, a Western democracy, an entire people—one with whom Americans have long had a special affinity and affection—had turned away from a major component of the disposable economy. They had plasticized a common product, and people had indeed come—and then they left it in the dust and found they felt better off for it. The convenience of the disposable bag had turned out to be either an illusion, or simply not worth the trouble. The ban has survived all complaints and court challenges since.

Other countries around the world soon adopted similar policies. Taiwan's 3-cent bag fee was a fraction of Ireland's, yet it knocked down plastic bag consumption 69 percent. Mainland China banned the bags outright, and consumption there dropped 66 percent. Bangladesh imposed a ban as well after severe floods were linked to storm drains clogged by plastic bags.

In 2005, the plastic bag battle moved back to the U.S. when San Francisco's Commission on the Environment voted unanimously to recommend that the city adopt a measure based on Ireland's law—a 17-cent fee to shoppers for every plastic or paper grocery bag. Proponents expected the fee would drive a reduction in disposable bag use—and pollution—just as occurred overseas. But there was one key difference. Ireland had no domestic plastic bag manufacturers to battle and lobby against a bag tax. America did.

The mayor and the San Francisco Board of Supervisors decided to move cautiously, postponing action in order to study the actual costs of plastic bag waste and to try out what proved to be an ineffectual voluntary program with grocery store chains to track and reduce plastic bag waste. The delay proved fatal to the tax proposal, giving the plastic bag industry time to form a coalition with the motto "Sack the Tax," and to lobby successfully in the state capitol to block San Francisco's bag fees. The resulting legislation bore a green title—"The Plastic Bag and Litter Reduction Act"—but its most meaningful provision was to outlaw all local plastic bag fees. San Francisco—and every other city in California—would be permanently barred from adopting Ireland's bag tax. It was a huge victory for the plastic industry, one of their biggest ever.

San Francisco officials responded with their own end-around. They couldn't impose a tax, but they realized nothing in the new law forbade them from adopting a simple and outright local ban of single-use plastic bags.[6] In April 2007, San Francisco became the first city in America to outlaw the single-use plastic grocery bag. Only paper bags with at least 40 percent recycled content, compostable plastic bags or reusable bags were allowed.

The ordinance garnered extensive international press, energized environmental groups who saw this as the first chink in the plastic industry's long record of victories over such measures, and helped propel San Francisco to the top of innumerable lists of green cities.

But in the years that passed, it's become clear that the ban, while on the surface more final and far-reaching than an Ireland-style bag tax, has proven to be the less effective method of the two approaches—at least if the goal is to do the most possible to curtail waste, save energy and cut plastic bag consumption and pollution. The tax approach also provides the most freedom of choice to consumers. In Ireland, you can always just pony up the 22 euro cents if you really have to have that plastic bag. By contrast, the San Francisco ban was fragmented and compromised by design. For one thing, it applied only to stores with $2 million or more in annual sales—basically the city's large supermarkets, retailers and drugstores. Hundreds of smaller shops continued to hand out free plastic shopping bags. So while the city reported a 50 percent reduction in plastic bag litter on the streets, The Pit at Recology is still awash with the things. Unlike Ireland, which forbade retailers from substituting free paper bags for taxed plastics, San Francisco supermarkets and drugstores could give away all the free paper bags that customers wanted. All told, the San Francisco ban did not usher in the same sort of sweeping changes in consumer behavior that Ireland pulled off with its bag tax, which had made the most environmentally sound choice—reusable bags—king of the checkout line.

Nevertheless, San Francisco's success at imposing any sort of plastic bag regulation despite heavy opposition from the plastic industry set off a wave of copycat proposals in communities throughout the state, as well as elsewhere in the nation, that has continued to this day.

The plastic industry began a campaign of lawsuits to stem this tide. The manufacturers reacted too slowly to mount a timely legal challenge to San Francisco's ordinance, but across the bay, financially strapped Oakland had to abandon its ban when plastic manufacturers sued to force the city to conduct an expensive environmental impact report on the ban to determine what *negative* ecological effects banning plastic bags would have. What would it mean for the environment, the plastic bag makers asked, if people turn to paper bags instead? What about the forests? What about climate change? It was a brilliant strategy, as the makers of a major source of litter and pollution made from fossil fuels turned environmental laws against the environmentalists. The plastic makers chalked up a series of victories against other proposed bans throughout California, as well as one proposed in Seattle.

But a dozen other communities, from the small Southern California coastal town of Manhattan Beach to the entire sprawling county of Los Angeles to Aspen, Colorado, to the Hamptons in New York and Westport, Connecticut, imposed plastic bag bans and fees anyway. Los Angeles County, which was sued by plastic makers in 2011 over its ordinance, tried to address the shortcomings of the San Francisco law by both banning plastic bags and imposing a 10-cent fee on paper grocery bags (to be collected and kept by retailers so it couldn't be branded a tax). In a significant setback for plastic bag makers, the California Supreme Court then ruled in 2011 to reject the tactics used against Oakland—requiring a costly environmental impact report for every community seeking a bag ban—as a violation of "common sense." The decision came in the case of little Manhattan Beach in Southern California, but the ruling provided a lawsuit-resistant template for any other community in the state to pursue a plastic bag ban.

Three thousand miles away, New York City contemplated a bag tax in 2010 but backed off in the face of industry opposition. That same year, Washington, D.C., imposed a nickel fee on both paper and plastic bags, with revenues going to clean up the litter-choked Anacostia River. Plastic bag consumption immediately dropped from 22.5 million bags a month to 3 million. Retailers agreed to give out hundreds of thousands of free reusable bags, which helped city council members resist a heavy lobbying effort against the measure.

Plastics lobbyists spent even more in Sacramento—more than $2 million—desperate to fight a statewide ban of single-use plastic bags in the nation's most populous and plastic-consuming state.[7] This was the bill that Andy Keller and his Bag Monster army had championed as the best way to deal with plastic bag pollution—with one consistent state law, rather than a host of communities duking it out in court one at a time with plastic bag manufacturers and imposing different conditions on retailers from town to town. Keller campaigned for the bill by decrying the wastefulness and environmental harm of plastic bags as well as the tactics of "Big Plastic," which he contrasted with the green jobs and sustainable products that could be created in California if reusable grocery bags became the mainstay of shoppers in the state.

"Groups like the American Chemistry Council are spending millions of dollars to stop this movement," he said at a Sacramento press conference, with then-governor Arnold Schwarzenegger at his side. "I think they would be elated if we all ran around as bag monsters."

But the plastic makers prevailed. Though it passed the state assembly, the bag ban died on a 14-21 vote in the state senate; each of the no voters had received campaign donations from the plastic industry. California's disposable economy remained intact.

The battle lines shifted then back to individual cities, where the disposable plastic bag was very slowly losing ground. Three leading plastic bag companies decided it was time to expand the industry's legal strategy and stretch those battle lines beyond activist city councils. They launched a new lawsuit to ensnare a certain reusable shopping bag company, and a CEO with a penchant for mocking "Big Plastic" by wearing five hundred disposable grocery bags.

CHICOBAG ISN'T the only green enterprise to build advocacy into its business model and then take heat for it. TerraCycle, a New Jersey company that has become a leader in "upcycling," faced a similar, potentially fatal attack from a larger, richer, established rival just as it was getting traction in the marketplace. Its experience would provide a model for Keller as he struggled to survive what he now calls "The Plastic Bag Wars."

TerraCycle had started out in 2001 as a business contest entry by two Princeton University freshmen—Tom Szaky (future chief executive officer) and cofounder Jon Beyer (future chief information officer). They made a deal with the university dining hall to take food waste for a worm farm that would turn Ivy League table scraps, plate scrapings, banana peels and coffee grounds into rich fertilizer. They would package the resulting "worm castings" in used soda bottles, making it the ultimate eco-product—fertilizer made from waste, packaged in waste. The students would even ship it in old, used cartons scrounged wherever they could be found, reasoning that reusing was infinitely preferable to recycling for any footprint you could think of: energy, carbon, water, forests, resources. The TerraCycle label informed potential customers that the key ingredient was "liquefied worm poop." The words "worm" and "poop" were in very large and colorful capital letters—truth in advertising, along with a whiff of potty humor. Szaky was, after all, still a teenager. The proposal placed fourth in the contest.

Convinced his idea was better than that outcome suggested (and, indeed, TerraCycle ended up winning several other contests that helped finance the start-up), Szaky borrowed, charged and emptied his college savings account to bring his worm poop to market. With a staff of mostly unpaid college students living in an old house in a dodgy Trenton neighborhood (Princeton being too pricey), TerraCycle made its first shipments of bottled worm poop to small nurseries and garden shops, and struggled to stay afloat. When Home Depot decided to carry TerraCycle's product in 2004 and later promoted it as an eco-friendly choice, the red ink started turning black. Walmart and other big boxes followed, and TerraCycle began expanding its product line into more varieties of wormy plant foods and fertilizers, which it promoted as completely organic and superior to the more commercial, chemical-based products that had long dominated the marketplace. Then TerraCycle started a "Bottle Brigade" program in which fundraising groups nationwide could fill TerraCycle-provided mailing boxes with twenty-ounce soda bottles, which were then shipped back to the company and used to bottle worm poop. Brigades earned donations to the schools of their choice for every bottle collected. The varying sizes and shapes of the bottles became part of the TerraCycle look, and the brigades concept would become a key part of their business strategy.

By 2006, the company had reached a million dollars in annual sales and Szaky was smiling from the cover of *Inc.* magazine, next to a headline reading: "The Coolest Little Start-up in America. Meet TerraCycle, the ultimate growth company. Built on garbage. Run by a kid. Loved by investors."

The federal lawsuit hit the next year, in March 2007. The established leader in the plant food business, The Scotts Company, makers of the Miracle-Gro line of products, alleged in its 177-page complaint that TerraCycle was copying the bigger company's distinctive yellow-and-green boxes—what's known as a "trade dress" violation—as well as breaking the World War II–era Lanham Act, which bans false advertising claims that harm another brand's business. The Lanham Act is the go-to law for big companies upset with the marketing claims of upstarts, as they are complex and expensive cases to litigate—which gives the Goliaths of the business world the advantage. TerraCycle that year was on track to crack $2 million in sales. Miracle-Gro was a global business with $2.7 *billion* in annual sales, by far the largest brand of lawn and garden products for consumers and professionals in the world, not only for its own products, but also with an exclusive contract to sell the Monsanto Company's popular home weed-killer, Roundup.

The company had long been in the crosshairs of environmentalists who decry its promotion of toxic herbicides, pesticides and fertilizers—a category of chemicals that can disrupt soil ecosystems and leach into waterways, harming wildlife. Miracle-Gro's trademarked "four-step program" for beautiful lawns is really a recipe for converting grass to a toxic "chemical dependency," according to a coalition of twenty-eight environmental and health organizations.[8]

By contrast, TerraCycle's claims about the relative virtues of organic versus synthetic plant food seemed mild. But the Miracle-Gro makers took exception to TerraCycle's claims that scientific research showed that its organic product was superior to the leading "synthetic" product. The bigger company also complained about TerraCycle's use of the term "goof-proof"—a phrase that Scotts claimed as a trademark—and for asserting that its organic product will not "burn" plants if over-applied, as chemical fertilizers may do. The lawsuit demanded that TerraCycle destroy its allegedly copycat labels and signs, stop comparing itself to Miracle-Gro and hand over all income it earned from the alleged imitation and denigration of Miracle-Gro products. In other words, it was time for the kiddies to close up shop and go home. The start-up could never survive if the behemoth prevailed on these demands.

TerraCycle hired a lawyer it couldn't afford to file the requisite briefs denying all the allegations. But the company's real weapon had nothing to do with the courts. Szaky fought back on the Internet with a website dedicated to the case, SuedByScotts.com. The site pioneered the now common tactic of launching litigation websites as a line of defense that—with a little Web savvy, cleverness and social media skills—can put even a houseful of college students on a more even footing with a well-funded corporate power.

On its pages, TerraCycle argued that the makers of Miracle-Gro were going after a business created by college students because it wanted to keep its near-monopoly powers in a market where it already controlled at least 60 percent of the business. In addition to posting copies of the legal documents in the case for all to see, TerraCycle put up a photographic comparison of eighty different plant food and fertilizer products from other manufacturers that also used the yellow-and-green color scheme Miracle-Gro claimed was so distinctive. The montage seemed to suggest that, first of all, the green and yellow colors were a virtual industry standard, and second, that of all the products in the lineup, TerraCycle's worm poop soda bottles were the least likely to be confused with the staid, rectangular Miracle-Gro items.

A collection of unflattering media statements from Jim Hagedorn, the CEO of Scotts, was tucked into the website, too: "What I like about this company is we kind of said ten years ago that we're going to take over the world. And we did . . . I kind of want to be at war all the time, and people aren't always comfortable with that."

The website shamelessly played up the David vs. Goliath aspect of the case, comparing photos of TerraCycle's modest Trenton headquarters (featuring a view through a torn chain-link fence) with the massively landscaped Scotts headquarters, with its pillared entrance and sparkling fountain. The site also contrasted the executive benefits each company CEO enjoyed ("unlimited free worm poop" for Szaky, "personal use of company-owned aircraft" for the Miracle-Gro leader). Under chief executive background, Szaky was identified as a "25-year-old Hungarian-born refugee college dropout," while Scotts's CEO was described as a "51-year-old former jet pilot, son of multi-millionaire Miracle-Gro Founder." A chart put Miracle-Gro's annual profits at $132.7 million; the profits column for TerraCycle simply said: "Not yet."

As a fundraising tool for TerraCycle's legal defense fund, the website proved to be a dud, pulling in a mere $515 after three months. But as public relations, it put TerraCycle on the map as never before; Szaky had correctly guessed the suit would provide the company's best marketing opportunity since its founding. Major articles on the company and the case appeared in the *Wall Street Journal*, the *New York Times* and hundreds of blogs and websites. Even in the *Journal*, the bible of big business, Miracle-Gro came off as a bully. Traffic to TerraCycle's website jumped from one thousand visitors a day to thirteen thousand, while sales revenues increased 122 percent in the weeks following the launch of the SuedByScotts .com website.

The case settled within six months. TerraCycle agreed to stop claiming to be superior to Miracle-Gro and to alter its packaging color scheme once existing stocks ran out, to stop using the phrase "goof-proof" and to take down the SuedByScotts.com website. No money changed hands, and each side paid its own litigation costs.

The legal victory, then, went to Miracle-Gro, which made no concessions. The benefits, however, all went to TerraCycle, which considered the packaging changes minor but the publicity it got in the process priceless. By 2008, the company was solidly profitable; revenues reached $7.5 million in 2009 and $20 million in 2010.

Scotts's Miracle-Gro, meanwhile, saw a reversal of its fortunes in subsequent years, as droughts and recession pummeled the lawn care businesses, its stock dropped precipitously, and the EPA concluded in 2008 that Scotts had sold four pesticide products that were "unregistered, falsely and misleadingly labeled, or both."[9] All four products were ordered recalled.

Meanwhile, TerraCycle achieved rapid growth as a green company by moving the company's products beyond the worm poop lineup by extending its original vision for rethinking and reinventing waste. Szaky expanded the company's bottle brigade program to collect all sorts of consumer product waste that was difficult or impossible to recycle—chip bags, energy bar wrappers, yogurt cups, juice pouches. Then TerraCycle scientists and engineers found a way to "upcycle" these environmentally damaging waste products. The company did not melt, crush or pulp them—the standard, energy-consuming recycling tactics—but just cleaned them up and found novel ways to construct products with them. He got the companies that made the products—Frito-Lay, Cliff Bars, Capri-Sun—to pay for the collection program, and then he took their waste, with their highly visible brands still on them, and turned them into Capri-Sun backpacks, Doritos jackets, clipboards, notebooks and a host of other products (the coasters and clipboards made out of old circuit boards are particularly cool). These were then sold in Target, Walmart and other major retailers. Szaky called his new business model "sponsored waste."

By 2011, sponsored brigades in fourteen countries were sending billions of pieces of trash to TerraCycle—a vast, green supply chain of upcyclable garbage—and in return, the brigades got up to 2 cents per piece for their school or charity. It's a business model that has made TerraCycle one of the fastest-growing green businesses on the planet—and it might never have happened without a big lawsuit to propel a worm-poop farmer into the big time.

LIKE THE CEO of TerraCycle, Andy Keller can laugh these days about his legal battle with a more powerful rival. Now he can cheerfully suggest that, in the end, his small company benefited from being targeted by what he thinks of as "Big Plastic"—even though the plastic companies who sued him also claim victory.

"They can say that," Keller says, "but I think it's clear this lawsuit was the biggest mistake they ever made. It put ChicoBag on the map. It gave me a national, maybe a global stage, to talk about these

issues, to talk about kicking the disposable habit, being less wasteful, protecting the environment—and how each of us can do our
part. And it ended up costing us nothing. We didn't have to do anything we wouldn't have done anyway."

But at the time of the filing, Keller wasn't laughing. The suit—
with its accusations that ChicoBag had engineered "a continu-
ous and systematic campaign of false advertising and unfair
competition"—had the potential to ruin his still fledgling green
business.

The suit was filed by a trio of plastic bag makers—Hilex Poly
Company of Hartsville, South Carolina (ten plants in seven states),
Superbag Operating, Ltd., of Houston, and Advance Polybag, Inc.,
of Sugar Land, Texas. Together they are among the largest plastic
grocery bag makers in the country. All three are members of the
Progressive Bag Affiliates of the American Chemistry Council (an
elite group; fellow members were: The Dow Chemical Company,
ExxonMobil Corporation, Total Petrochemicals USA, and Unistar
Plastics). Hilex Poly was also heavily invested in the Save the Plastic Bag Coalition, a front group with a single employee—a lawyer
whose full-time job is to sue (or threaten to sue) any California city
that attempts to ban plastic bags.

The three bag companies asserted that ChicoBag used its "Learn
the Facts" company Web page to cause the plastic bag makers "ir-
reparable harm" through deceptive trade practices and false claims.
There were five statements the plastic makers took issue with that
had been published on ChicoBag's website:

- A reusable bag needs only to be used eleven times to have a
  lower environmental impact than using eleven disposable bags.
- Only 1 percent of plastic bags are recycled.

- Somewhere between 500 billion and 1 trillion plastic bags are consumed worldwide each year.
- The world's largest landfill can be found floating between Hawaii and San Francisco. Wind and sea currents carry marine debris from all over the world to what is now known as the Great Pacific Garbage Patch. This "landfill" is estimated to be twice the size of Texas and contain thousands of pounds of our discarded trash, mostly plastics.
- Each year hundreds of thousands of seabirds and marine life die from ingestible plastics mistaken for food.

The plastic companies did not explain how such general claims published on a reusable bag company's website could affect their commercial disposable bag business. The vast majority of consumers had never heard of Hilex Poly or the other companies. Proving the five statements false would not win the case for the plastic makers; they would also have to show that these statements caused specific injuries to their business and income—a formidable legal hurdle. They would have to prove, for instance, that executives at a major supermarket chain perused ChicoBag's website and then decided to stop buying plastic bags from one of the three defendants. That's the sort of specific damage that has to be proven under the Lanham Act, the same legal bludgeon Miracle-Gro used against TerraCycle—you don't win just by alleging some vague injury to a company's reputation. And that specific damage would have to be linked directly to ChicoBag, as opposed to the hundreds of other websites and major media outlets that had similar, and often more damning, statements about plastic bags causing environmental harm. What this suit really was about, one legal commentator opined at the time, was targeting a single outspoken critic of plastic bags in hopes of shutting up other critics who might not want to be next on the firing line. ChicoBag would be an example, an object lesson—a deterrent.[10]

The suit was filed in South Carolina, which has nothing like California's law against SLAPP suits (Strategic Lawsuit Against Public Participation). In his home state Keller could challenge the validity of such a lawsuit as a veiled tactic to censor, intimidate and silence a critic by saddling ChicoBag with backbreaking legal bills. In South Carolina, that wasn't an option. The two sides would have to fight the case on its merits.

ChicoBag has several factors in its favor. It's common for websites to report statistics, even controversial and extravagant ones, without much (or any) sourcing. ChicoBag, however, had annotated its "Learn the Facts" page and provided the sources for the five statements challenged by the plastic bag makers. The sources were reputable ones: the EPA, *National Geographic* and the *Los Angeles Times*. And the ChicoBag website had quoted these sources correctly—that wasn't in dispute. But if the claims were false and they actually harmed the plastic bag business, even repeating them accurately could still constitute a violation of the Lanham Act.

The plastic bag companies first let their displeasure with ChicoBag be known by sending a cease and desist letter to Keller, a common prelude to suing that gives alleged miscreants a chance to clean up their acts before being taken to court. And Keller responded to the letter by taking those five statements off his site until he could investigate and, if there were errors, correct them. He adds, "Then they sued me anyway."

But Keller still conducted his investigation—and decided it was the plastic companies, not ChicoBag, who had it wrong.

For the first statement targeted in the lawsuit, Keller discovered the EPA's website no longer displayed the information he had relied on about shopping with a reusable bag eleven times. That didn't make the information false, just inaccessible. So he turned to a life-cycle report on supermarket bags that Hilex Poly itself had cited. That study confirmed Keller's point: that a reusable grocery bag made of non-woven polypropylene plastic would have to be used at least eleven times to have a lower carbon footprint than using disposable single-use grocery bags. There were other comparisons in the study, too: Using a paper bag three times would do the trick, while it would take 131 trips to the market with a cotton bag to have a lower carbon footprint—which meant the material used for a reusable bag was critical. The cotton footprint sounds very high, but given the average American's five-hundred-bag annual habit, it would still be an improvement given the long life of a well-made cotton bag. In preparing for his legal defense, Keller hired a scientist who had worked extensively for Hilex Poly and asked him to perform a life-cycle analysis. The scientists found that ChicoBag's line of products made out of recycled PET plastic (basically old soda bottles) did even better than Keller's website had claimed: One of those bags had to be used only *nine* times to have a better environmental footprint than disposable bags.[11]

As for the second claim—that only 1 percent of plastic bags were recycled—there's no question that the EPA reported this statistic in its 2005 Municipal Solid Waste Report. The plastic bag makers admitted as much, but complained that the information was so dated as to be misleading. But 2005 is the last year such a statistic on the recycling of plastic bags is even available from the EPA. After that year—due to the plastic industry itself—the statistics for bags were combined with plastic wraps and films. Keller repeatedly requested separate plastic bag stats from Hilex Poly, but none ever came.

Combining the recycling stats from different kinds of plastic is what's really misleading, Keller argues. But even using those more recent combined numbers, the recycling rate for plastic films, bags and wraps is still less than 10 percent in more recent EPA data. "Anemic," Keller calls it. "Certainly nothing to brag about."

The third challenged statistic—that the worldwide consumption of disposable plastic bags of all types is 500 billion to 1 trillion—may not be accurate, Keller realized after investigating further. But that's only because the number is almost certainly *higher*. Keller compiled plastic bag consumption data for the European Union, China, Australia, Japan, Canada, India and the U.S., and painstakingly detailed on his website the sources and calculations that led to the 1 trillion annual plastic bag consumption estimate. That figure does not include areas for which he lacked good data: South America, Africa, Central America, the Middle East, Eastern Europe and a sizable portion of Asia—which means, Keller says, that the trillion estimate is in no way unfairly criticizing the plastic industry. It's giving them a free ride.

Keller's description of the Pacific Garbage Patch as a giant floating landfill was based on reporting by the *Los Angeles Times* (which used the phrase "world's largest dump" in its Pulitzer Prize–winning "Altered Oceans" series of reports[12]). This is similar to a spate of news and blog reports on the gyre that likened the plastic pollution to a floating continent or a mass of debris. Using the term "landfill" or "dump" is more metaphorical than literal, Keller says, and there are definitely more precise ways to describe it. But his description, as he sees it, only underestimates the challenge posed by the huge area of ocean where plastic pollution is concentrated. It can lead people to envision a more solid, visible collection of plastic waste—and therefore an easier one to clean up—rather than the diffuse soup of small confetti-like particles that's really out there, and that he witnessed firsthand while accompanying a research voyage with the 5 Gyres Institute.

Finally there was the fifth challenged statement, which stems from one of the most oft-repeated but thinly sourced statistics on marine pollution—the death of hundreds of thousands of seabirds and marine life from marine plastic debris. Many publications and websites have repeated this information before and after Keller wrote his "Just the Facts" Web page, some of them advancing the claim that there are *millions* of seabirds killed every year by plastic, not hundreds of thousands as Keller's site stated. Keller's version is sufficiently general and conservative enough to be supported by the available research, according to the National Oceanic and Atmospheric Administration.[13]

Convinced that his original information, while it could be improved, was not false and misleading, Keller expanded his plastic research. He created and published on his website an extensive timeline on the many lawsuits and lobbying efforts by the plastic industry to preserve their disposable bag business and to undermine opponents, dating all the way back to the industry's successful battle a half century ago to block a ban of plastic dry-cleaning bags that had been linked to child asphyxiations.

The lawsuit and Keller's dogged response brought media attention and increased sales to ChicoBag. *Fortune*, *Rolling Stone*, the *New York Times*, the *San Francisco Chronicle* and other media outlets covered the case, and Keller relished the role of "little guy facing off the schoolyard bully."

"I really think they were trying to make an example of me," Keller later reflected. "The plastic bag is under attack all over the world, and after years of Big Plastic winning all its battles, people and communities finally began waking up to the fact that it's crazy to make a product that gets thrown away after one use, that lasts for hundreds of years, that gets blown away and washed down rivers and into the ocean. And then here I come with Bag Monster, and I think it really got under their skin. It was the last straw for them."

Keller pinned his legal defense on nine scientific experts on ocean plastic pollution, marine plastic ingestion, life cycles of plastic bags and the plastic bag industry. According to Keller, their reports showed that, if anything, ChicoBag's page of facts about plastic bags had only underestimated the true potential for environmental harm and costs.

Before a trial could settle the issues, and before even the deposition stage of the lawsuit, in which Keller and the plastic bag executives would have to submit to hours of questioning, Hilex Poly decided to settle. The other two companies had already dropped out of the case.

The settlement imposed conditions on both sides. Keller agreed to revise his facts page in order to attribute statistics to current reports and data rather than any archived or outdated reports. He also agreed to use the newer plastic bag recycling figures that combine bags with wraps and film, as well as reporting that the last recycling figure reported strictly for plastic bags was 1 percent. Hilex Poly likewise agreed to stick with the most current and clearly sourced statistics on its own website, and to make clear the nature of the mixed statistic on bags, wraps and films. The company also agreed to put the statement "Tie Bag in Knot Before Disposal" on its bags, and to discuss on its website how this simple precaution when throwing away plastic bags can prevent windblown litter.

ChicoBag's insurer agreed to make a cash payment to Hilex

Poly, the amount of which was kept secret under the terms of the settlement.

Both sides claimed victory. Hilex Poly's press release on the settlement was muted and factual, emphasizing that the settlement would ensure a fair and open debate on plastic bags in the future, calling it "a win for consumers." Its wording carefully stopped short of stating that the allegation of false statements by ChicoBag had been proved; instead, a vice president of the company was quoted as saying, "The use of false and misleading statements is injurious to the marketplace, and this settlement ensures that facts are accurate."[14] But a press release from the other two plastic bag companies, Superbag and Advance Polybag, companies that dropped out of the case before it settled, bordered on the bizarre. Among other things, it accused ChicoBag of creating "an imitation EPA website to share false information," something that was never alleged in the suit, much less admitted or proven.[15] This, according to Keller, was the very definition of false and misleading statements.

His own press release portrayed the lawsuit as "frivolous" and the settlement as a big win for his company and the environment, particularly the bag maker's agreement to depart from its historical stance that "Bags don't litter, people do." Instead, Hilex Poly acknowledged that bags can become windblown litter even when properly disposed of. Telling consumers to tie their bags in knots was a huge shift and concession, Keller says, as was Hilex Poly's agreement not to mislead the public with combined statistics that could make plastic bag recycling rates look better than they really are.

"What started as a bullying tactic," his press release said, "has morphed into two wins for the environment."

————

WHETHER THEY are the crinkly white sacks at the local supermarket, the clear sandwich bags in kids' lunch bags, the flimsy baglets that protect newspapers even on sunny days or the bags for the carrots and celery in the produce aisle, plastic bags are ubiquitous. Frozen ravioli, cotton balls, socks, potatoes, jelly beans, pinto beans—you could fill a book with the items that come to us encased in plastic bags. The average American touches plastic bags multiple times a day, hundreds of bags a year, many thousands in a lifetime. Even when, as their makers argue, disposable bags serve a second purpose at home—as a trash bag or a pooper-scooper or to wrap a school project in on a wet day—most still end up in a landfill soon after they enter the home. Others find their way into storm drains, rivers, oceans. They are mostly not recycled despite decades of efforts. Even their biodegradable counterparts rarely make it to a facility that can actually recycle them.[16] Their environmental footprint and cost are greater than the simple expedient of a reusable bag. They are, as Andy Keller is quick to point out, a product with a useful life measured in hours and a waste life measured in centuries.

That said, plastic bags are a comparatively modest part of the waste stream. They are part of the marine pollution problem, but how much remains unclear. They take up room at landfills, but other packaging forms a bigger part of the 102-ton legacy. Why, then, are so many cities making bags a priority? Why is Andy Keller so passionate about it that he would put his company and livelihood at risk rather than abandon his efforts to undermine single-use plastic bags and persuade others to give them up for good?

Because, Keller says, as a symbol, few parts of our waste stream and our disposable plastic economy are more potent and visible in our daily lives. And few parts of the 102-ton legacy are easier for an ordinary person to change.

Keller believes that the single-use plastic bag habit—his bag monster—"is the poster child for unnecessary waste." Breaking the habit of being a bag monster, he says, is the first step in moving our homes, our families and our communities into less wasteful, more reusable habits and consumer behavior. First get rid of the bags, then move on to other disposables that we don't really need, says Keller. At ChicoBag, he ditched paper towels for a next step. Each employee was given a cloth towel with a hook to hang it on. It gets washed as often as necessary.

Keller next bought each employee a thermal container for drinks and a reusable clamshell for salads and sandwiches for casual meals and takeout at restaurants that usually serve on plastic, paper, foam or other single-use diningware. Restaurants that were willing to serve the zero-waste way got ChicoBag employees' business. Others that refused lost those customers, though the ChicoBag workers made it clear they'd happily return if the management reconsidered its position. When it dawned on restaurant owners that they were losing paying customers for no better reason than habit and old thinking, that it was no harder to serve food without wasting paper and plastic—and, in fact, it saved packaging costs—several changed their minds.

These kinds of incremental changes add up, Keller says, altering the dynamic of the consumer culture, because each one gets easier than the last. According to the ChicoBag founder, the way to start that particular snowball rolling is with the plastic bag. It's harder to accomplish without the obvious, clear benefit of the Irish-style bag tax—the plastic industry has for the most part fought off that sort of clear-cut incentive in America, dulling the message that consumers can simply skip the bag and save money. The industry also has a strong counter-message: Banning bags will cost jobs, fees will hurt the economy and consumer spending, and they'll spawn a new government bureaucracy. Just using the word "tax" is a potent weapon. Never mind that we're already paying an invisible bag tax, Keller says, because they're not really free—consumers pay for them in the form of higher food prices at the market, about $30 a year per person. It's not as if retailers are going to pay $4 billion a year for bags and not pass on the cost to customers. The lost jobs and economic harm arguments were raised decades ago by the paper industry in hopes of staving off competition from plastic. Then, as now, it was an instance of an entrenched but aging business model—paper—losing its revenues to the upstart—plastic. The paper bag companies complained of lost jobs, but what they were really fighting was the *shift* of jobs (and profits) to the plastic newcomers. One person's job killer is another's progress. Now plastic bag makers are marshaling the same old arguments, this time fighting a shift from the disposable economy to a reusable one. They have even become champions of recycling, which they initially resisted, because it's a way to keep consumption rates high for disposable things. "The myth of recycling," says Keller, "is that it's okay to consume all we want as long as it has that little recycling symbol on it. But that's completely false, and just perpetuates our wasteful, disposable ways."

Keller talks about plastic bags and the disposable economy in terms of addiction. For him, the cure starts with the bag, because it has to start somewhere.

"Bags are kind of like the gateway drug to all the plastics," Keller says, "and if we can kick that habit, all the rest of our single-use habits will start to fall like dominoes."

PLASTIC BAG RESTRICTIONS, U.S JURISDICTIONS, BY YEAR OF ADOPTION

All entries are plastic bag bans unless bag fee is noted.

* indicates bag ban approved but not yet effective or enjoined by lawsuit

2007   San Francisco, CA

2008   Malibu, CA
Fairfax, CA
Manhattan Beach, CA
Westport, CT
Maui County, HI
Seattle, WA (overturned by lawsuit)

2009   Fairbanks, AK (5-cent bag fee rescinded by town
            council after one month)
Palo Alto, CA
Kaua'i, HI
Edmonds, WA
Westport, CT

2010   Los Angeles County, CA
San Jose, CA
Telluride, CO
Maui, HI
Brownsville, TX
Washington, DC (5-cent bag fee)

2011   Santa Monica, CA
Calabasas, CA
Long Beach, CA

Marin County, CA*
Monterey, CA
Pasadena, CA
Santa Cruz County, CA
Santa Clara County, CA
Sunnyvale, CA
Aspen, CO
Portland, OR
Bellingham, WA
Seattle, WA

2012

State of Hawaii (first statewide ban)
Alameda County, CA*
Carmel-by-the-Sea, CA*
Carpenteria, CA
Dana Point, CA*
Fort Bragg, CA
Laguna Beach, CA*
Los Angeles (city)*
Mendocino County, CA
Millbrae, CA
Ojai, CA*
San Luis Obispo, CA*
San Francisco Part II—extended to include all retailers
    and takeout food
Santa Cruz (city), CA*
Solana Beach, CA*
Ukiah, CA
Watsonville, CA*
West Hollywood, CA
Corvalis, OR

At least 55 other communities had bans pending or under study by the end of 2012.

GREEN CITIES AND
GARBAGE DEATH RAYS

THERE'S ONE CITY IN AMERICA THAT CONSISTENTLY rates at or near the top of every list and survey of sustainability, green buildings, recycling rates, clean transportation, energy efficiency and eco businesses. This city pioneered smart growth policies back in the 1970s and kept them in place through the deregulation fervor of the 1980s. This required considerable upstream swimming. The Reaganesque mantra had taken root elsewhere: *Government's the problem, not the solution.* In most communities, that philosophy had led to an era of unprecedented private exploitation of public lands, of endless strip mall construction, of minimal oversight and of widespread hostility to the environmental protections set in place a decade earlier.

But as a result of its contrarian leaders and citizenry, this city ended up with ample green space, the largest wilderness park within city limits in America, walkable neighborhoods and a rich web of local farms that supply a famously locavore restaurant and farmers' market scene. Yet it still has an urban core sufficiently prosperous that financial analysts have called it the "comeback city" for its job growth even during the recession.

This same city is so obsessed with green and zero-emissions transportation that some businesses offer more parking for bikes than cars, bicycle lanes are everywhere, many city intersections have green zones that allow bikes to cut in front of car traffic, and it's a major destination for the increasingly popular "car-free vacation." Fitness clubs use exercise bicycles and their own members to generate green power. The city's public transit doesn't stop with trolleys, trains and buses but also includes an aerial tramway that looks like it was beamed in from an Alpine chalet—a cable car that brings commuters up from the new south waterfront downtown development to a hilltop university campus and medical center, the city's largest employer. It carried a million passengers by its tenth month in service, in a town with fewer than six hundred thousand residents. This city even adopted a climate change plan in 1993, five years before the famous Kyoto Protocols, and was the first (and only) American city to meet the Kyoto goals years ahead of schedule by reducing its greenhouse gas emissions below 1990 levels beginning in 2008. It did this despite growing at a faster rate than most other cities in the country, providing a counterpoint to the conventional wisdom that communities can have strong environmental protection or economic growth, but not both.

In fact, prosperous Portland, Oregon, does so many green things right that its greener-than-thou sensibilities have spawned a sardonic cable TV show, *Portlandia*, which features such gems as a mayor who bicycles directly into his office at City Hall and dismounts onto a giant inflated ball that doubles as his desk chair. Of course, the real mayor, Sam Adams, who appears on the TV show as the fictional mayor's aide, isn't so different. Out in the real world, he has sung on the radio, "Bring your bag, bring your bag, bring your bag, bag, bag . . ." to the tune of the *William Tell Overture*—as a reminder to Portlanders to bring their reusable shopping bags to the grocery store.

There's just one area where sustainable Portland lags, the big challenge any community with green aspirations must wrestle and beat: its trash.

They make a lot of it in Portland—a shade more trash even than the average American's 7.1 pounds a day, and a half pound more than the average Oregonian. The last time per capita waste statistics were released, residents of the Greater Portland Metropolitan Area—just "Metro," as it's referred to locally—generated more than 1.3 tons of trash a year. That's 7.14 pounds a day per Portlander.

They do a good job of diverting much of that trash from the landfill—with about 59 percent recycled, composted or burned for energy in 2010 (that's for the entire three-county, 2.3-million-person Metro area; within the city limits, the 586,000 urban residents of Portland do even better, hitting 67 percent). That's still behind San Francisco's official 77 percent rate, but well ahead of the national average of 24 percent. Yet even with all that landfill diversion, Metro Portland still sends sixty massive trucks every day laden with garbage to the Columbia Ridge Landfill in Arlington, on the border with Washington. That's a truck of trash every twenty-four minutes, setting out on a 360-mile round-trip to the landfill and back. For a town so proud of its fleet of LEED-certified (Leadership in Energy and Environmental Design) sustainable buildings, and with a Port of Portland headquarters that is also a living water and waste treatment plant—a kind of man-made wetlands inside an office building—the trash issue is a painful shortcoming. Diesel trucks hauling garbage long distances, then depositing the trash in landfills—a major source of greenhouse gases—is a model Portland leaders want to change, which is why they began debating in 2011 what the future of trash would look like in 2020, the year current contracts for waste hauling and disposal lapse.

Future Portland may feature ramped-up composting plants, or generate electricity through anaerobic digesters—vats that speed up the decomposition of garbage, then use the resulting methane to make electricity or vehicle fuel. The trash futurists are also anticipating greater recycling rates and reductions in disposable plastic and paper consumption, perhaps by pushing for product stewardship rules under which manufacturers would have to take responsibility for the waste their products leave behind. This is a nascent movement at present, but some companies—Patagonia, the clothing and outdoor equipment maker, has been a leader in this—tell customers to send their purchases back for reuse or recycling when they are done with them, no questions asked. The question is, can a community encourage such a business model? Mandate it? If anyone would be willing to give it a go, it's Portland.

Meanwhile, at the other end of the waste-management spectrum, a test facility is coming online at Portland's primary landfill destination in Arlington to study the effectiveness of the experimental waste-treatment process known as plasma gasification—a technology that vaporizes garbage with arcs of electrical energy that heat matter inside their beam to 25,000 degrees. This is not burning trash. Indeed, the process takes place in the absence of oxygen, and so many of the normal, noxious byproducts of combustion are not produced. The process yields a synthetic gaseous fuel and a lump of shiny rock, not unlike volcanic glass, with toxins locked up inside in relative safety. This garbage death ray reduces trash volume by 99 percent, not even leaving ash behind. Just a hunk of obsidian, about twenty pounds' worth for every ton of trash disintegrated.

Scaled up, if such a technology proves cost-effective, it could make big landfills obsolete. But it is the longest of long shots, and not just because the technology at the moment is prohibitively expensive. Getting energy from trash remains exceedingly unpopular among American environmentalists. It has a long and dirty history, marked by the heated sorts of battles that upended California's big plans for a landfill-free future in the 1980s. New York City remains scarred by similar battles. Although the technology and its pollution controls have advanced since then, old objections and distrust remain. The Sierra Club, among other groups, adamantly opposes attempts to ramp up trash-to-energy projects, and that carries weight, especially in Portland.

There is also the question of reducing waste that none of these end-game strategies address—all the burning, landfilling, recycling and composting does is redirect our 102-ton legacy. How does a town like Portland stop making so much garbage in the first place? Like so many communities across America, Portland is not yet sure what magic mix of technology, technique, inducements, prohibitions and exhortations to consumers to change their behavior should be attempted in the hope of actually reducing the 102 tons we are destined to leave behind, rather than merely shuffling it to some other form of treatment. But uncertainty or not, the deadline to decide is approaching.

"We will have the next evolution in waste in place before 2020," says Matt Korot, Metro's director of resource conservation and recycling. "We know we can't wait until the last minute. We're just not sure yet what that's going to look like."

NOW CONSIDER another city. It, too, is routinely listed as one of the greenest cities on the planet, and also one of the most livable. Its parks are legendary, rich with history and plentiful, and they're being aggressively expanded so that, by 2015, every resident will be within a fifteen-minute walk of park or beach. A world-beating 40 percent of workers commute each day by bicycle, from bankers in business suits to factory workers in hard hats. Workplace culture puts the CEO and the mailroom clerk and everyone in between on a first-name basis, allowing bonds and unity within companies that can be tough to match elsewhere in the world. This city's central river and canals, once polluted, are now safe for swimming, a feat that earned a prestigious international environmental award in 2000. It is also the organic food capital of the world—45 percent of food purchased there is natural and chemical free. It is closing in on a goal of 90 percent organic food served in school cafeterias and retirement homes.

This city has led its entire country from foreign oil dependence to energy independence over the past three decades. It is on course to use zero fossil fuels by 2050. Since 1980, it has reduced energy consumption (and global warming emissions, though that was not the initial goal) while doubling its economy and offering a standard of living, health care (free to all), education (ditto) and amenities that match or exceed the best the U.S. has to offer. On the downside, taxes and energy costs are higher than in even the most expensive U.S. city. Yet polls of residents show a majority feels these burdens are more than offset by the absence of medical, insurance and tuition bills; by a more conservation-conscious culture when it comes to purchases, energy and fuel; and by the far lower incidence of crime, hunger and poverty than U.S. citizens experience. That's worth some extra taxes, Peter Bach, a civil engineer for the national energy department, told the *Wall Street Journal*, echoing the sentiments of a majority of his countrymen. The *Journal* found itself writing admiringly[1] about this country's energy independence and conservation-embedded lifestyle, despite the fact that its success at achieving what has eluded America essentially defies every principle Wall Street holds dear. "You can't just sit back and wait for markets to do this for you," Bach told the financial newspaper.

On the garbage front, this city is so far ahead of its American counterparts that it's like comparing laser surgery to leech craft. This city recycles trash at twice the U.S. average, its residents create less than half the household waste per capita, and the community philosophy holds that dealing with and solving the problem of trash must be a local concern, even a neighborhood concern. When it comes to waste, NIMBY (Not in My Backyard) is not a factor, as shipping trash off to some distant landfill—making it disappear for others to manage—is considered wasteful, costly and immoral. Not that such out-of-sight, out-of-mind garbage treatment is much of a consideration here: only 3 to 4 percent of this city's waste ends up in landfills, compared to the U.S. average of 69 percent.

This is not some Shangri-la of past or future. It is the Copenhagen, Denmark, of today. And the secret sauce for that city and the entire nation of Denmark, at least on the waste disposal front, is its mastery of turning trash into a renewable energy source.

"They are the model, along with Japan and a number of other countries in Europe," says Nickolas Themelis of Columbia University, America's engineer-apostle of the untapped power of garbage. "They put these waste-to-energy plants right in their neighborhoods. They become part of the fabric of the community. There's none of the fear and misinformation about waste energy that we have in the U.S. They are clean and efficient, and many of them are quite attractive. The people are *proud* of them."

Denmark's strategy has been to build trash-burning, power-generating plants on a relatively small scale. No behemoths burning 2,000, 5,000 or 10,000 tons of garbage a day, such as those proposed for Los Angeles in the seventies and eighties, only to be shot down by concerns over pollution and neighborhood impact. Instead, the Danes built a network of community-based plants that average in the 400- to 500-ton-a-day range throughout their small nation of 5.5 million inhabitants. The largest handles about 1,000 tons a day— the Amagerforbrænding plant on the outskirts of Copenhagen, dating back to the 1970s (and upgraded many times since, primarily with added layers of emission controls). Urban neighborhoods, suburban enclaves, upscale areas and working-class housing all are served by these plants. Keeping them local eliminates the cost and the emissions of having to haul trash long distances across the city or countryside, as often occurs in the U.S., where trash travels millions of miles every year just to get from municipal transfer stations (like The Pit in San Francisco) to landfills. Another benefit of the local Danish plants: They not only generate electricity in place of coal-fired power, they also pump heat through a vast network of underground pipes to keep houses and businesses warm, thereby doubling the efficiency of the plants while taking the place of less efficient home furnaces. Some American city centers use this type of heating, often called cogeneration or "district heating"—New York and Denver among them, where the systems date back to the 1880s—as do a number of large university campuses (notably the University of New Hampshire uses landfill gas to make all of its heat and energy). But Denmark has expanded the concept to the point where more than six out of ten Danish homes are heated this way. The system is credited for half of Denmark's energy savings in the past quarter century. The larger of these waste-to-energy plants can generate up to 25 megawatts of electricity (enough for fifty thousand households) and district heating for 120,000 or more homes.[2]

The push to build such plants, along with a heightened commitment to bicycle-friendly policies and an advanced public transit system, began with the oil crisis of the 1970s, when Arab oil-producing nations embargoed countries that supported Israel. Gas lines, rationing, economic upheaval and inflation resulted. Like many nations at the time, Denmark launched initiatives to develop renewable power and energy independence so it could never again be blackmailed or coaxed to take sides by foreign oil suppliers. When the political situation changed and the oil started flowing freely again, most countries, none more than the U.S., quickly abandoned government stimulus for renewable energy and aggressive mandates for conservation and auto fuel efficiency. But a few countries, Japan and Denmark chief among them, decided it would serve multiple purposes—national security, economic stability and environmental protection—to stay the course on key elements of those programs. The climate, the global economy and the politics of energy would be in a very different and certainly less dire place today if Denmark's approach had been the majority view rather than the minority's.

Since that time, twenty-nine waste-to-energy plants were constructed in Denmark; as of 2011, ten more were in the works, planned or already under construction. The subterranean heating systems required a massive public works undertaking, with extensive and disruptive excavations that took years. But when it was done and the hot air started blowing, home heating bills in the cold Scandinavian climate dropped to a fifth of what they had been. Rebates and tax incentives accompanied a government mandate for developers, homeowners and businesses to thoroughly insulate buildings to avoid wasting this new heating energy; a similar set of incentives led to the mass purchases of energy-efficient appliances, with adoption rates reaching 90 percent.

At the time of the seventies oil crisis, Denmark depended on foreign fossil fuel supplies for 90 percent of its energy. Now it is energy independent and a net oil exporter from its modest offshore drilling operations.

One of the largest wind power installations in the world was constructed on the coast, and Denmark became a world leader in wind energy with the rise of its homegrown company Vestas, maker of state-of-the-art wind turbines. The industry was jump-started in the seventies with major investments of research dollars, loan guarantees and subsidies from the government; now little Denmark owns a third of the multibillion-dollar global wind turbine business. This is a business America initially dominated before the post-oil-crisis renewable energy incentives were killed.

Nineteen percent of Denmark's power is generated by wind, the highest in the world. An even greater amount of the country's energy supply is derived from trash.

In a country where electricity is generated by nonprofit electrical cooperatives in which the ratepayers are also the plant owners, attitudes about utilities and energy are very different from those in America. In Denmark, polls consistently show that a majority of Danes are willing to pay a higher rate if the electricity is clean and produced with domestic fuels.

The waste-to-energy system as it exists in Denmark today uses a tried-and-true technology called "mass burn," which aptly describes how the facility operates. Trash trucks deliver garbage to the plants, with recyclables already separated out. The garbage is then pushed into furnaces by a series of moving grates. The burning trash heats boilers to create high-pressure steam. A flue gas cleaning system and banks of filters remove pollutants so thoroughly that the very tall smokestacks that used to be the main feature of such plants—so pollution would be dispersed far above neighborhoods—are no longer necessary. The output of the most harmful byproducts of incineration, including the main environmental showstopper in the U.S., dioxins, has been reduced to levels that represent a mere fraction of what the average home fireplace or backyard barbecue puts out. These plants are now so clean that they exceed European pollution standards (generally stricter than in the U.S.) by a factor of ten, and the trash-based power is considered a form of renewable energy. The plants not only emit less greenhouse gases than coal plants, they are also superior to landfills in that respect, where even the most advanced methane capture systems (such as Puente Hills's landfill gas power station) still allow 50 percent of the climate-busting methane to bleed into the atmosphere. Methane has twenty-three times the global-warming punch as the carbon dioxide produced by combustion. The Denmark waste-to-energy plants scour and sift the solid residue of the incineration process, called "slag," removing metals for recycling as well as other useful chemical byproducts. What's left is sold to construction companies to use for concrete and road building. Nothing is wasted.

Denmark and other countries in Europe, where there are more than four hundred waste-to-energy plants in service, have made many of the structures architecturally attractive, incorporating art, sculpture and novel design. Some are community centers, some anchor parks. The Spittelau plant in Vienna, with its brightly colored design and illuminated globe perched atop a tower, has become a tourist destination. A garbage power plant was built in Paris that burns 1,260 tons of trash a day less than three miles from the Eiffel Tower; surrounded by trees and topped with a grassy, living roof, it is all but invisible. A new state-of-the-art plant is now under construction in Copenhagen to replace the original aged Amagerforbrænding plant by 2016. The new facility will double as a community ski park, the tall incinerator building serving as the anchor for three separate slopes of varying difficulty while, beneath the snow, the trash from five municipalities will be burned to make power and heat for 140,000 homes. The Danish architect Bjarke Ingels has designed the facility with a chimney that will blow smoke rings each time it accumulates a quarter ton of carbon dioxide from the burning trash. The idea is to be playful while also reminding people that their consumption has consequences, that trash power still exacts a price on the environment, and that the best strategy for dealing with waste is to waste less.

"They are so far ahead of us," Themelis says. "Our behavior in the U.S. in this area is really quite irrational. And it's irresponsible. We are throwing energy and money away every day, burying it in the ground."

Themelis has been researching and advocating trash energy for decades now, building an international network of experts and engineers by founding the Waste-to-Energy Research and Technology Council, with branches in the U.S., Greece, Germany, Japan, India, Brazil, Mexico and China, where the government has been on a waste-to-energy building spree since the turn of the century. A native of Athens, Greece, Themelis directs the Earth Engineering Center at Columbia's Fu Foundation School of Engineering and Applied Science. A chemical engineer by training, his initial work was in the private sector, where he developed a more efficient method of copper smelting that drastically reduced sulfur emissions in the mining industry. When he came to Columbia University in 1980, he arrived at what was then called the Henry Krumb School of Mines, where he says the historical emphasis was the "three Ms"—mining, materials and metallurgy.

Eventually Themelis helped lead a group of faculty members who shifted the emphasis to the "three Es"—earth, environment and engineering—which led to renaming the school the Earth Engineering Center. Waste-energy research has dominated his career ever since. He says the current system of burying waste in landfills amounts to burying a billion barrels of oil a year that could be used for much needed energy.

Themelis frets that the same arguments against waste-to-energy used in the eighties are still being used to keep the technology limited in the United States, which has eighty-seven waste-to-energy plants, almost all of them dating back to the early 1990s or before. Even though the emissions controls have advanced and more than satisfied tough European environmental standards, fear over dioxins and other pollution remains great and these are often still cited by opponents. Yet a 2009 study concluded that harmful emissions from landfills were greater than those from modern waste-to-energy plants.[3]

"There's just a great deal of fear about it," says Themelis. "It's like some isolated tribe who has never seen an airplane before, and is frightened of it. They just close their eyes to it."

Aside from concerns over emissions, which proponents (as well as European environmental agencies) assert have been solved by new technology, a principal argument against ramping up waste-to-energy in the U.S. lies in its poor energy and economic bottom line when compared to recycling. The plants really are expensive—with large-scale facilities costing in the $600 million range and up—and trash is not a very good fuel, so the power output per dollar spent on boilers and generators is less than in a conventional power plant. Recycling trash, on the other hand, has a lower environmental impact and, pound for pound, can save more energy than burning the same trash produces. Recycling aluminum cans, for instance, saves a whopping 96 percent of the energy needed to produce aluminum from bauxite ore. Recycling glass jars and bottles saves 21 percent of the energy needed to make new glass, recycling newsprint saves 45 percent, and recycling plastic beverage bottles saves 76 percent (other plastic types differ in the percentages, but the energy savings are there, too). More recycling, then, is a better strategy than waste-to-energy, Themelis's detractors say. Critics of the technology also fear that adding trash-burning plants to the mix will discourage recycling because expensive plants will demand more trash in order to pay off their hefty costs.

Themelis says that recycling's energy advantage is real and that it often is the better alternative, but not always. There is a flaw in the recycling case: After a certain point, the energy gains are more theoretical than practical for many types of trash. Those theoretical energy savings are often not realized because recycling some materials still costs more than using new raw materials. Recycling plastic grocery bags, for example, costs four to five times what the raw materials are worth. Transportation costs, manpower for sorting recyclables from garbage and contamination problems make recycling a lot of common items of trash too costly or too difficult or both, despite the energy savings. This is why recycling rates, outside of a few highly committed U.S. cities, are far lower than the amount of recyclable material in the American waste stream. And it's why even America's recycling leader, San Francisco, still sends trash to the landfill in which two-thirds of the material is theoretically recyclable. This is why the king of trash says he's got ten billion dollars in value locked up in the trash he hauls to the landfill, if only he could tap into it. He would if he could. But that capability has eluded us.

Themelis argues that material, then, should be used to make energy, not garbage mountains. This would not hurt recycling, he suggests, but would augment it.

Furthermore, Themelis says, no recycling process is 100 percent. There is always 10 to 15 percent residue left behind that can't be recycled. No one ever counts that or subtracts it from the total amount recycled in those cities with such high recycling numbers, Themelis points out. All that material is counted as "diverted," then the leftover muck just gets quietly carted to the landfill. But that residue also would be perfect fuel for a waste-to-energy plant.

The final argument against waste-to-energy—that it will reverse gains in recycling—is belied by the history of the technology, Themelis points out. The cities and nations that have made trash burning a key part of their energy and waste strategies—Denmark, Germany, Austria, Japan, the Netherlands—all have robust recycling programs that not only recycle as much as or more than the amount of trash that is burned, but they all also recycle at a much higher percentage than the U.S. has been able to accomplish. It's the landfilling that diminishes when waste-to-energy becomes a strong option, not recycling. Germany, for instance, burns 34 percent of its municipal waste and it recycles the rest, an impressive 66 percent. That's not just one super-green city, like San Francisco, but an entire country of 82 million people, the powerhouse economy of Europe. Almost none of its municipal waste gets landfilled.

Waste-to-energy opponents also base their negative views on the assumption that the only choice for garbage power lies in very large, expensive, utility-scale trash power plants—which, in fairness, is the only type seriously attempted in the U.S. But the most successful use of the waste-to-energy technology right now has been the smaller, less costly, community-based plants that Denmark and other European nations favor—a more distributed power generation system rather than the central utility style used in the States.

In an odd parallel, this same focus on utility-scale power plants, with their huge upfront costs and requirements for immense transmission lines, has similarly handicapped solar power development in the U.S. Other countries have focused on distributed, rooftop solar power, which does not require huge capital investment or immense transmission line upgrades. All it takes is a law compelling utilities to pay a market rate for home-brewed solar power. Germany has used this approach—called a "feed-in tariff"—to become a world leader in solar power generation, even though it has far less overall sunshine than the U.S. landscape. American utilities, meanwhile, have successfully lobbied against many such measures to boost home-based and small-scale solar, even as they cut deals based on government incentives, such as access to cheap federal land in the California desert, for large-scale solar projects that have yet to make a dent in our coal-and-oil-dependent economy.

Waste-to-energy advocates argue in favor of embracing a more distributed model for trash power in the U.S., which could work just as well in American cities as in Denmark. Steven Cohen, director of Columbia University's Earth Institute, has suggested New York City try such an approach, with fifty-nine small plants, one for each community district in the city. These plants could combine waste-to-energy with recycling and anaerobic digestion for composting organic waste. It would be a far better investment, he says, than spending $300 million a year to truck garbage out of state—an investment that has nothing to show for it at the end of a year other than an immense legacy of diesel emissions. But New York politicians have been burned so many times by trash burning that no one wants to even talk about it anymore.

Although it has garnered little attention nationally, it turns out the U.S. actually has a couple of mini-Denmarks of its own already. Lee County, Florida, where the city of Fort Myers is located, in 2007 added 50 percent more capacity to an existing garbage power plant, then closed down its last landfill. The county now recycles or composts half its trash, and burns the rest to make enough electricity to power thirty-six thousand homes.

Even more impressive is the state of Connecticut, America's garbage power leader with six waste-to-energy plants, most built in the 1980s. They handle 62 percent of the state's trash, supplying about 10 percent of the state's electricity needs. Twenty-six percent of Connecticut's waste is recycled, with about 12 percent sent to landfills, the lowest of any state. (Connecticut residents also make less trash than the national average—about five and a half pounds a day each—a lifetime trash legacy of 78 tons.)

The Connecticut program has been so successful that the state is scheduled to close its last active landfill by 2015. The plants have paid for themselves many times over.

And yet the head of the Connecticut Resource Recovery Authority says that, such success notwithstanding, they could never get another such plant built in today's political and economic climate.

FOR ALL his advocacy for waste-to-energy, Nickolas Themelis believes that the most intelligent, most-likely-to-succeed, long-term solution to waste is far simpler than any giant trash-burning generator, and far less costly, yet so much more difficult to achieve: a changed culture.

He believes there must be a shift to a culture that wastes less, one that demands products that are less wasteful, and that embraces products designed to "close the loop"—to be reborn and remanufactured, rather than thrown away or burned. Call it "cradle to cradle," architect and environmentalist William McDonough's catchphrase. Call it zero waste. Call it conservation. Call it liberal, progressive or conservative, to Themelis it doesn't matter. The point is, he says, there needs to be a change. And someday, that change will come—not out of choice, he says rather gloomily, but out of necessity.

"But then there's what can be done now. Today. And the reality today is that we have to work with what we have, and with where we are *now*. And right now, we're burying treasure instead of using it to power our homes. And that's shameful."

Could a Portland emulate Copenhagen on trash, much as it did years ago when exploring ways to make itself more bicycle friendly? Should it try? Should it even consider the waste-to-energy question, or continue to pursue ever-more recycling in the hope of achieving something close to zero waste—the goal San Francisco has set for itself? And is that goal even possible?

Themelis says no. For that matter, so does Andy Keller of ChicoBag, who sees recycling as a crutch to allow Americans to feel better about overconsumption of disposable things. How does a Denmark, with a robust economy and a standard of living at least as good as America's, manage to make half as much trash per person as the U.S.? Keller suggests that consumers anywhere can make a culture shift all on their own, by looking at the kind of things being purchased and asking: *Is this thing I'm buying going to be in the trash in a year or two? Or is that going to be useful and treasured for many years to come?*

"If you're buying something and thinking it could be an heirloom someday," Keller says, "then you're on the right track."

These men who made a study of waste central to their very different careers ended up focusing on the same solution in the end: a societal shift from the culture of disposable abundance to a more measured consumption, a focus on quality over quantity, on more carefully chosen treasures. It sounds great in theory. Getting there is another story. A hint of what that future might look like may be in Denmark, which has shifted more toward sustainability than the United States. But even some communities in America have made inroads of their own. Plastic bag bans—removing a wasteful object, rather than redirecting it to some new destination—has become the first baby step toward lowering the 102-ton legacy in a growing number of American cities.

Metro Portland's waste and recycling czar, Matt Korot, says all of these ideas could find their way to the table as one of America's greenest cities plots the future of waste. Given the long commute Portland's trash currently takes, a Denmark-style shift to trash heat and power would seem to be an attractive alternative. So far, it's not been discussed beyond the small plasma gasification experiment out at Portland's remote landfill, a far riskier, more exotic and un- proven technology compared to the tried-and-true mass burn trash reactors now in place all over Europe and in Connecticut.

For now, it's clear the momentum and the desire in green cities such as Portland are with recycling and composting. And even that can be a tough sell. In 2011 Portland adopted a San Francisco–style plastic grocery bag ban that brought complaints from all sides. There were those who missed having the bags for trash and dog poop-scooping. Others saw so many loopholes to the ban that it seemed next to useless, as exemptions included bags for produce, meat and bulk food at groceries, as well as vendors at the popular Portland Farmers Market.

Then there's the new food scraps pickup process, in which Port- land hopes to catch up with San Francisco. After a year-long pilot study, Portland launched its household food waste collection service to shrink the landfill loads and divert the food part to composting. Every home received an official pail with a lid for the kitchen coun- ter to hold smelly food wastes in until they could be dumped into the curbside bin.

The well-thought-out plan had a rub, however: The new weekly food scraps collection meant other trash services had to be cut. Regular garbage would be picked up every two weeks instead of weekly. Considerable civic grumbling ensued. People were upset about having to keep out pails in the kitchen and overflowing trash bins in the yard.

"What's easier," the mayor quipped in a newspaper column, clearly irritated with his normally green constituency, "cutting gas- oline use by three million gallons a year or getting Portlanders to toss pizza crusts into a pail on the counter? If Portland food scraps stay out of landfill where they produce greenhouse gases as they decompose, then we can keep up to thirty thousand tons of carbon-equivalent emissions out of the atmosphere in a year."

Given these difficulties, plans to expand the food waste pickups beyond the city limits to the entire Portland Metro area may take years as residents debate that balance between convenience and environment. If selling composting is so difficult in Portland, waste-to-energy might be a nonstarter. And if it can't fly in Portland, where can it fly?

"We need recycling, as much of it as we can do," Themelis says. "And for now we need landfills. But the missing part of the puzzle is waste-to-energy. Hopefully, we'll wake up . . . We can't wait until the whole culture changes to a less wasteful one. We must act now."

12 PUT-DOWNS, PICKUPS AND THE POWER OF NO

BEA JOHNSON NEVER SAW HERSELF, HER HOME, HER family or her habits as particularly wasteful. Certainly no more so than any other family with two active kids and a big house in the San Francisco Bay Area suburbs, where driving was the only viable option for getting anywhere. They put the recyclables in the right bin. They shopped at farmers' markets so they could buy local foods. They did their part, as Johnson had seen it. What could be wasteful about that?

Then a move to a new town landed them in an apartment while they searched for a new house, and Johnson realized just how wrong she had been. The new apartment was less than half the size of their old three-thousand-square-foot home, and so everything but the bare essentials—including a lot of bulky furniture, extra clothes, a garage full of boxes of who knew what—went into storage.

It didn't take Bea Johnson very long to realize something: She missed exactly none of it. She didn't miss those two extra sets of dishes, the extra sets of silverware or the "good" wineglasses. Not the whole rack of clothes that had filled her closet but that she almost never wore. Not the extra shoes, including pairs she hated or that hurt her feet, but she never seemed to part with. Certainly not those chairs that had looked nice in all that extra space in the old place, but that no one actually wanted to sit in. She didn't miss the clutter, or the gadgets on the kitchen counter, or the multiple TV sets. All that stuff she had spent years buying and accumulating, all the packaging and boxes and shopping bags and money that went with it, was gone, and instead of missing it all, she discovered that she reveled in its absence.

That's when Bea Johnson finally got it: There's power in putting things down instead of putting them in your shopping cart. There's power in saying *no*—the power to change a family's life and fortune. Maybe a community's. Maybe a whole country's.

"I like it like this," she told her husband one morning. In years past, there would have been messes throughout the house to clean up—clothes and toys and extra dirty dishes in the sink that would rob hours from her day. Now five minutes of pickup and she'd be done. "I want to keep it like this."

That's how it started. Not with a conscious effort to be greener or more sustainable or less wasteful. The Johnsons didn't know about, hadn't thought about, the 102-ton legacy back then, and so they weren't formulating a strategy for eco-consciousness-raising.

They had just stumbled on the fact that they were happier in a simpler, less cluttered home, and agreed that they'd see where that idea would take them. They'd downsize a bit, cut out the impulse spending, the recreational shopping, and see. They had no idea this would lead to a near-zero-waste lifestyle, where their lifetime legacy of trash is on track to be measured in pounds rather than tons. These days, a year's worth of trash for the Johnson house—the stuff that can't be recycled, repurposed, given away or composted—fits in a mason jar.

"And we have never been happier," Johnson says.

Some people, even friends, are put off by the lovely but spartan home, by Johnson's indifference to shopping, her preference for thrift stores when shopping is unavoidable, her adamant resistance to anything packaged or plastic—basically, her lack of attachment to stuff, particularly the disposable stuff that drives our economy and fills our trash cans. "I could never live like this," one girlfriend flatly told her. "Why would I want to?"

Johnson smiled and said she wanted to show her something. The friend expected to be shown some literature on the evils of plastic, or toxic landfills, or the planetary benefits of sustainability—the usual green justification for crazy eco-hippie behavior that strays far from the American norm. Instead, Johnson pulled out a page of figures her husband, a business consultant, had penciled out on what their new, low-waste lifestyle cost. She hadn't needed that sort of practical information to support her desire to simplify and downsize, but Scott had. "He figured we're saving about forty percent over what we used to spend," she told her friend.

A YEAR'S WORTH OF UNRECYCLABLE TRASH IN THE JOHNSON HOUSEHOLD

Receptacle: one large mason jar containing:

- several pieces of bubble gum
- plastic wrappers from prescription bottle
- plastic tamper-proof seals from contact lens fluid
- expired laminated ID card
- plastic stickers from grocery store fruit
- backing from postage stamps
- clothing tags (the itchy ones)
- masking tape from a paint job

Four out of ten dollars that they used to spend, in other words, were wasted on things they either didn't need or didn't want. That's made a huge difference for the Johnsons in a difficult economy. They can afford cool vacations. A new hybrid. A generous college fund for the boys.

Her dubious friend was quiet for a long beat. Then she said, "So how can I get started?"

BEA JOHNSON is a slim woman in her mid-thirties with long, dark blond hair with light streaks (achieved by adding a strong brew of chamomile tea for highlighting to her shampoo and conditioner— all bought in bulk, package-free, she is quick to point out. She is serious about living the low-waste life, which for her includes shunning packaging, harsh chemicals and plastic as much as possible— which isn't as hard as it sounds, she says, though you have to be able to say no with regularity. Her gravity and determination are leavened by an easy laugh, particularly when she's explaining the chemistry of making her own low-waste cleaners. (With vinegar and castile soap in hand, anything is possible, except—she has to laugh at this—for her disastrous attempt to concoct laundry detergent. Not a good idea, she chuckles, unless you want all your clothes gray.)

Johnson is an artist by training, though the gallery showings and artistic output have taken a backseat to her small business, Be Simple, which helps other people de-clutter and de-trash their homes and lives as she has done. She is also very French, her heritage and accent equally strong. She pronounces her name BAY-a; her younger son, Leo, is LAY-o. She first came to America as an au pair, fell in love with California and a young business consultant named Scott Johnson, and they married. They lived in Europe for four years, then returned to the States to raise their family. Bea says she was determined to live the American Dream, and she thought she knew exactly what that was supposed to look like. "It meant the big house, the walk-in closet, the SUV, the yard with the dog and the white-picket fence. And stuff. Lots of stuff. With a big garage filled with more stuff."

The Johnsons had that and then some. Bea became a recreational shopper. Friends would come to visit and they'd go shop for fun, because that's what you did. She'd walk into Target and come out with a bag of purchases and a hundred dollars more debt on her credit card and, a week later, have no idea what she spent it on. Keeping the house in order, the kitchen clean, the kids' clothes picked up all seemed like a full-time job. Beneath the kitchen sink there was a jungle of plastic bottles containing every type of polish and cleaner known to man—window cleaner, tile cleaner, floor cleaner, cleanser powder, hand soap, dish soap. The medicine chest looked like a shelf at the supermarket. The trash cans were full every week.

Then came the move. The family had lived for seven years in that big house in Pleasant Hill, but as the boys grew, Bea and Scott increasingly missed the ability to walk places that they had so enjoyed in France. Their community was designed for the car. So they chose a more walkable city, Mill Valley, just across the Golden Gate Bridge from San Francisco. They rented an apartment near the town center, and started looking for a house, with a majority of their things in storage. They ended up taking a year before they found the right home at the right price. By then, they had grown accustomed to living with less. The kids barely noticed. Scott was busy with a new business—he had quit his secure job with the regular paycheck and launched a start-up sustainability consultancy with three partners—and so he ceded household decisions to Bea. And she loved the new order. Everything was easier. There was less cleaning, less organizing. It was voluntary simplicity, and she persuaded her husband they should make it permanent. It was a good time to economize, she added, and this would be the perfect way to do so.

When they found a new house with about 1,200 square feet plus a small basement, that cinched matters. The furniture and boxes in storage weren't going to fit. They ended up selling 80 percent of it—the extra sets of china, the extra clothes, the gadgets and sporting goods nobody used. All gone. They furnished the house in clean, elegant whites, uncluttered countertops, largely unadorned walls and a sectional couch in the living room that could be separated into chairs, converted to a guest bed and reconfigured in a host of other ways. They put their habits under a magnifying glass, looking for ways to simplify and de-clutter. If they had been saving things that just sat on a shelf, untouched and dusty, out they went. Books that had been read but were not treasures were sold; the library would fill their reading needs.

Next came reading and research on the environmental consequences of waste—of plastics in the ocean, of consumer chemicals in the human body, of the rise of packaging as the number one source of trash, of recycling as more panacea than solution, because most plastics had only a few recycles in them before they reached an end state as trash. Bea began to focus on minimizing waste, looking for alternatives for things they bought and consumed that made trash. Was there a non-trash alternative? In most cases, she found, the answer was yes. That didn't mean giving up nice things or denying themselves. They weren't living in rags or eating granola. They were simply saying no to the disposable economy and everything that came with it. "We didn't become hippies," she laughs. "Living with less doesn't mean living poorly."

There was a learning curve to this process of zero-waste discovery, requiring experimentation, research and a false start here and there (like the failed attempt to make laundry detergent). But over time, the Johnsons have made their habits and their home a model of waste*less*ness. Not perfect, not absolute. But they have made an impressive assault on the 102-ton legacy.

Bea says it wasn't hard to identify consumerism—the accumulation of stuff—as the main engine of waste, disorganization and unnecessary expense. So she added a fourth "R" to the traditional 3 Rs of green living—reduce, reuse and recycle. The fourth R is "refuse," as in refusing offers of disposable goods, processed foods and other items that fall into the broad category she calls "crap." "Refuse, refuse, refuse," she says, almost turning it into a chant. "Just say no, no, no. Someone offers you a free pen at a conference, some knick-knack you'll just throw in a drawer, some piece of crap you don't need—say no. Refuse. Because every time you say yes, you are inviting more to be made. You have created demand for more waste. So we refuse all of that."

The same attitude carries over to purchases. More than 10 percent of the cost of things lies in the packaging. So the Johnsons refuse packaged goods, plastic bottles, single-use bags. They buy in bulk. Stainless steel pump dispensers hold shampoo and conditioner in the bathroom, filled with bulk purchases at the market. A glass jar holds the tooth powder Bea makes from baking soda and a bit of the herbal sweetener stevia—no plastic toothpaste tubes and caps for this zero-waste house. Scott uses old-school double-edged razor blades in a stainless steel razor; the blades come in a tiny box with virtually no packaging, and each one lasts six months. A bar of rich Turkish soap provides the lather for shaving instead of canned or tubed foam. An alum stone serves as a natural deodorant for the whole family. Vinegar and water with a touch of castile soap is the Johnsons' single household cleaner, used on floors, counters, walls, glass, bathrooms—everything. The bulk castile is put in reusable pumps in kitchen and bathroom, used to wash dishes, hands, pets. Washable microfiber cloths are used for cleaning—no paper towels. The family does buy toilet paper, but only the kind wrapped in paper—plastic film and wrap is public enemy number one in the low-waste home. You won't find any in the well-stocked pantry, either: The shelves are filled with mason jars—"the French type, with the lids attached—they're much better," Johnson says. The jars hold grains, pasta, snacks, cereal, cookies, beans—all the items of any well-stocked pantry, except none of it comes in boxes, cans or plastic containers. She has a hundred of these jars in varying sizes—they're dishwasher, freezer and refrigerator safe.

But it is different, a major deviation from the norm that can induce near states of shock in visitors. Friends of her older son, Max, were over one day looking for a snack. They opened the fully stocked pantry, saw no cardboard boxes, plastic or foil-wrapped cookies—none of the familiar disposable packages—and said, "You don't have any food." The shelves were full of mason jars, including a row filled with kid-friendly snacks. But they might as well have been invisible.

Bea grocery shops once a week. First she gets produce at the farmers' market, tucked into a reusable mesh bag, then she moves to the one local grocery store that offers a large selection of bulk goods, from pasta to peanut butter, cookies to couscous. She brings three reusable shopping totes with a number of mason jars in them. Her market weighs the empty jars, then she can use them to purchase cuts of meat, fish, cheese and salad bar items. Her standing order for ten loaves of French baguettes is waiting for her; she cuts them in half, puts them in a pillowcase, and stores the bread in her freezer, doling it out during the week, warming it in the oven. The bread can be safely frozen this way; Johnson says most people think it must be hermetically sealed in plastic to protect the flavor, but this simply is not true. Johnson buys a local dairy brand of milk that comes in returnable, reusable glass bottles. She keeps two big mason jars of flour on hand all the time for baking—quiche, pizza, cookies. Because the bulk foods are all on the periphery of the market, and she never ventures down the maze of center aisles where all the cans and processed food are shelved, she usually finishes her weekly shopping faster than most people.

She shops for clothes twice a year, fall and spring, and limits her wardrobe to seven shirts, two skirts, one pair of shorts, three pairs of pants, three sweaters, three dresses and six pairs of shoes (including slippers). The rest of the family has a similar clothing inventory. Secondhand and thrift shops are the first line of attack in the clothing hunt, reuse being high on the list of low-waste commandments. Bea was particularly proud of one spring expedition in which she restocked the whole family's wardrobe for forty bucks, including several Abercrombie T-shirts for the boys at $1 apiece.

There are frustrations: Junk mail can't be refused, so it has to be recycled, the least desirable of the 4 Rs. She also resents the big folders of school pictures that come home every year whether you want them or not. She has repeatedly complained about the paper and plasticized strips that come off the mailing envelopes for DVDs from Netflix, but the company hasn't responded. She has taken to slipping the pieces of trash back in mailers and returning it to the company for disposal.

In the end, the result of all this waste reduction is a freeing up of Bea's time.

"Most people, the first thing they say is, all this fighting waste must take up so much time and effort. It must be exhausting. And I say, just the opposite. I have more time than ever for family activities, for art, for whatever I like. I spend less time shopping, less time cleaning, less time picking up. I've never been happier or more relaxed."

There's also the question of sacrifice. Virtually everyone assumes that Bea and her family have made enormous sacrifices by giving up so much in order to become less wasteful. But is having only one set of dinner plates really a sacrifice? Does she really need more than six pairs of shoes? How much of a sacrifice is it to make her own mustard in her own containers, when she can make a year's supply in about fifteen minutes at a fraction of the cost of store-bought condiments? Is it a sacrifice to take five minutes to mix some water and vinegar and liquid soap together in a stainless steel spray bottle and use it for all her household cleaning (and yes, it works great, she says) rather than buying a half dozen disposable bottles of cleaner at fifty times the price? Refusing things—and, specifically, disposable things—should not be confused with sacrifice, she says. Once upon a time, it used to require a sacrifice to *buy* something. You saved up, you gave up things you might want just so you could put enough money aside to purchase something big or long-lasting or vital. Now, she says, people tend to think the sacrifice is *not* buying. That's one reason we are swimming in waste, Bea says. In her view, not buying is never a sacrifice. It's a way of saving up for something really important, or saving time, or saving the planet. Or all three.

Her whole family has taken part in these efforts with enthusiasm, though the boys sometimes grumble about the limits on toy accumulation. Each has two bins, and all toys—Legos, trucks, whatever—must fit in them. If that means parting with something to make room for incoming, so be it. Her older son, Max, has nevertheless embraced the zero-waste philosophy with aplomb, including demonstrating for his fifth-grade class—and in a YouTube video—a Japanese cloth-folding technique for making a waste-free lunch. He makes a sandwich to bring to school, then does an intricate fold that turns a cloth napkin into a multi-compartment tube that holds the sandwich and a piece of fruit or a cookie, and rolls everything up into a compact parcel. At lunch, the cloth becomes his place mat during eating, and his napkin afterward. Then he puts it into his backpack to be taken home and reused or washed as needed. No paper, no wrap, no baggies, no trash.

For birthdays and Christmas, the family emphasizes the giving of experiences rather than things. Trips and outings, hikes, camping, movies, museums, amusement parks—things the family can do together, relish and remember. They have a living tree they haul in every year for Christmas and decorate with heirloom ornaments.

But Bea says their wasteless lifestyle is neither fanatical nor absolute. The boys desperately wanted a Wii video game system and the Johnsons, after much debate, said yes. Now Max wants a cell phone and his own computer, which his parents are resisting for the moment, despite the dreaded "all the other kids have them" argument. Battling the multimillion-dollar advertising budget of the consumer economy with the word "refuse" is and always will be a challenge, Bea says.

After about two years into the low-waste lifestyle, Bea started a blog about her experiences, "The Zero Waste Home," which offered both a narrative on her family's quest and tips for others interested in trying their own hands at the "4 Rs." The blog, in turn, generated news media interest, and eventually *Sunset* magazine showed up in Mill Valley to write about Bea Johnson's efforts and to create a photo spread and report on her home as part of a series entitled "Inspired Homes of the West."

The article drew tremendous reader interest—and some strong emotions. Six months after the article appeared online, comments were still flowing in, nearly seven hundred of them. Many people were inspired. They loved the Johnson home's smooth surfaces and open space, and the main work of art in the living room—a "living wall" of plants, green and lush. One reader admiringly called the magazine photos "minimalist porn."

But a sizable minority of comments were dismissive, critical or just plain mean. Bea was attacked for eating beef and therefore supporting a wasteful and environmentally damaging industry. She was pilloried for an annual trip to visit family in France because she didn't adhere to her zero-waste practices there, and left behind an enormous carbon footprint since air travel is one of the most wasteful of human activities. She was criticized for choosing compostable toothbrushes made in Australia—their carbon footprint is huge, one commenter sneered. (Responded Bea: Her choice isn't perfect, but still better than plastic toothbrushes, especially since almost all of them are made in China with an equally big transportation footprint.) There was an undercurrent of anger in those comments, as if the Johnsons' lifestyle constituted an attack on the commenters' choices and values. Some called her mentally ill, delusional, a child abuser (what, no junk food?) or an obsessive-compulsive. Their home and lives were deemed barren and stripped of creativity. They were fanatics who would destroy the country if their ideas were to spread. Singled out for derision was the family's practice of digitally photographing the kids' artwork, then recycling the paper copies. In all, the comments were so numerous and strong that one university professor left a note saying he was assigning his psychology class to study the responses.

"Look, we're not perfect," Johnson says, bemused but not surprised by the reaction. She got a similar, if more muted, array of admiration and incredulity from friends and family. "We eat meat. I like makeup, I like to look nice, it makes me happy. And some of it has packaging. So no, we're not perfect. But we're not attacking anyone. We're attacking waste. I believe the solution to our environmental problems starts at home. Not in Washington or on Wall Street, but in our individual buying choices. That's how we vote, with our dollars, and they can change things. So that's what we're doing. And isn't it interesting that it makes so many people angry? They seem to think I'm attacking the American Dream.

"But I'm *living* the American Dream. I used to think it was the big house and the SUV and all that, but it's not. It never was. That's not the real dream. The real American Dream is having financial freedom. It's being able to do what we want to do. It's saving instead of wasting. What's wrong with that?"

CAN AN "ordinary" person make a difference? That's the question Bea Johnson gets all the time. The problems are so big, people say, what difference can one person or one family possibly make? The Johnsons' answer to this classic (and frequently paralyzing) question was to attend to their own nest first, to put down the wasteful things that once defined their home and possessions, and then find ways to help others to do the same with Bea's blog and de-cluttering business, with their son's ninja-lunch YouTube videos, with the example of a different way they provide to friends and local businesses in Mill Valley.

There's no shortage of parallel examples of one person making a big impact. Rob Gogan, associate manager of recycling and waste services at Harvard University, took an epic instance of waste and turned it into a shining example of repurposing and reuse. Every year, when Harvard's students depart campus for the summer, they leave behind roomfuls of perfectly good couches, chairs, tables, lamps and all manner of household items, abandoned without a care. And every year, Harvard would clear it all out and throw it all away—until Gogan launched what has been billed as one of the largest yard sales in America, right in Harvard Yard. Instead of heading to a landfill, those rescued items draw buyers from all over the region in search of a bargain.

"It's a dream job," says Gogan. "Harvard is the wealthiest university in the world and affluence produces effluence. It's a rich vein of ore for a recycler."

Then there's Kim Masoner. She's on a parallel journey against waste, though unlike Bea Johnson's, hers started outside her home, and involves picking up waste rather than putting it down. But she, too, is testing the conventional wisdom that waste and pollution are too big for any one person's efforts to matter. Over the past ten years, Masoner has become the queen of California beach cleanups.

She draws an army of volunteers to the plastic-impregnated sands of Southern California, not only from the beach towns themselves, but from inland cities whose storm drains have sent the trash downstream to the coast. She is a diminutive, energetic, youthful fifty-something woman who's been obsessed with picking up litter since age seven, and there's something of the Pied Piper in her ability to get the unlikeliest characters to show up to help her pluck Styrofoam out of rocky jetties and plastic straws out of the dunes.

"It's our trash," the mayor of the inland city of Azusa explained when he showed up at one cleanup with two busloads of high school students eager to clean the sands of the sleepy town where Masoner lives, Seal Beach. "We want to help clean it up."

Masoner got started in 1999 when she and her husband, Steve, started bringing a trash bag along during their daily walks on the beach, picking up debris as they went. When other people began asking the Masoners if they had extra bags to share so that they could join in, an impromptu beach cleanup ensued. It snowballed from there. Now the former director of the Seal Beach Chamber of Commerce spends all her time running Save Our Beach, a nonprofit community group that stages the most heavily attended monthly beach cleanups in Southern California, routinely drawing a thousand volunteers at a time. Corporations have her stage custom cleanups attended by their executives. Then she leads them in a class to learn how to crochet used plastic bags into yoga mats and bedrolls for the homeless—some of the myriad ways that Masoner has learned to repurpose waste instead of sending it to the landfill. The next day she'll be teaching high schoolers how to make purses out of old videotape and bracelets out of trash. Her volunteers have picked up more than 200 tons of plastic debris and other garbage from Southern California beaches.

"We can tell we're making a difference," she says. "There's less trash every year."

Gogan and Masoner have been widely praised and admired for their efforts against waste. Masoner has received numerous awards. The Los Angeles Lakers honored her at a halftime ceremony. She is a beloved local hero.

The universally positive way in which these two are perceived makes for a fascinating contrast with the reactions that Bea Johnson provokes in people. The reasons for this are subtle but instructive. The first two trash-fighters identify a problem of waste in the outside world and ask people to give of their money or time to help solve it. And people do just that. They can spend money at a yard sale or spend time on the beach and help save the world—without making any fundamental changes in their own homes or lives.

But Johnson and her zero-waste crusade are a whole different animal. She has identified a problem not on a campus or a beach but inside everyone's home and lifestyle. And her family has responded by transforming itself in a dramatic way, becoming happier and more prosperous by rejecting the consumer economy and lifestyle most Americans live and breathe. Is there any wonder why this angers so many people? Agreeing with the Johnsons' views means you either have to accept living a wasteful life, or change. A kind of cultural physics comes into play in this sort of situation, a fundamental, almost Newtonian principle that states it's always easier to oppose change than to propose it. Or put another way, picking up trash on the beach makes us feel good. Admitting we lead wasteful lives that need to change—not so much.

BEA JOHNSON remains optimistic (except for infuriating influxes of junk mail, which she is still helpless to stop). Despite the anger and the fear of change so many people seem to feel about her ideas, she feels certain that the advantages of a low-waste life will catch on— if for no other reason than that 40 percent savings on the household budget she can attest to. Who can't use that?

She says she sees signs of progress. The slow rise in the acceptance of reusable shopping bags is a great first step. And the responsiveness of businesses she has urged to improve also encourages her. When she found disposable plastic tasting spoons strewn on the sidewalk downtown, she asked the nearby ice cream store to do something about it. After a brief boycott and the shop owner's tearful response, the shop began offering stainless steel tasting spoons that it washed between customers. Score another one against the disposable economy, Johnson laughs.

Someday people will realize what her family now knows: that the real sacrifice is clinging to waste, and that the real American Dream, the original version of it, is waiting for those who give it up.

"All of a sudden you become aware. And you say to yourself, what the hell was I thinking all those years? And that's a beautiful moment. That's where it starts."

BEA JOHNSON'S TEN WAYS TO GET STARTED ON THE LOW-WASTE PATH

1. Bring glass jars, totes, cloth bags and cartons to the grocery store to carry food.
2. Buy in bulk. It eliminates packaging and can be more economical in the long run.
3. Refill clean empty wine bottles at local wine bottling events instead of buying new ones.
4. Use microfiber cloths instead of paper towels.
5. Make your own multipurpose cleaner out of vinegar, water and castile soap.
6. Use handkerchiefs instead of paper tissues.
7. When buying makeup products, choose a company that takes its packaging back and recycles it.
8. Only recycle paper if it's been printed on both sides. Otherwise, use the blank side for making lists or jotting down notes.
9. Use cloth napkins instead of paper napkins. That means cocktail napkins, too.
10. When packing a lunch, wrap sandwiches or other food in a cloth napkin instead of using wax paper, plastic wrap or plastic bags.

EPILOGUE

GARBAGE IN, GARBAGE OUT

When this book was conceived, I intended to write about our *64-ton* lifetime trash legacy, not the 102 tons it turns out to be. This original, smaller calculation was based on the widely accepted and official data point produced by the U.S. Environmental Protection Agency, which asserts that the average American produced 4.5 pounds of trash a day. When I discovered midway through this project that these numbers were wrong, that Americans were actually churning out an average of 7.1 pounds a day and sending twice as much trash to the landfill as we were being led to believe, it did more than change the central metaphor of a book about garbage.

It meant our trash problem—our trash addiction—already the biggest on the planet, is way, way worse than we've been told. It also meant our solutions have been much more paltry than most people understand. We've barely put a dent in our collective garbage mountain, and what we have accomplished—moderately increased recycling compared to decades past—is more about rearranging the deck chairs than changing the course of a ship headed for disaster.

It would be easy to focus on the implications of these flawed numbers as a scandal or a crisis—or as evidence that there is little that can be done about those hundreds of millions of tons of trash getting shuffled to the curb and hauled off for burial. It would be so tempting to throw up our hands, to say it's just too big to confront, to surrender to trashy inertia. Really, what can any one of us do about an ocean, a mountain, a 102-ton leviathan of trash?

In a word: plenty.

The discovery that even our top garbologists can't keep track of our trash, that they are as clueless as the rest of us, is best viewed not as a crisis, but as an invitation and an opportunity—an opportunity to take a step back and consider a new normal. What better time than today, a time of economic hardship, to reconsider our wastefulness, to absorb the lessons of MIT trash trackers and ocean plastic netters and a family of four in Marin County, and then find our way back?

Think about it. We are wasting so much stuff every day that trash has become a geographic feature—particles transforming oceans, garbage mountains dominating landscapes, a landfill visible even from space. How can that be acceptable? How can that have become normal?

Or look closer to home and see just how our daily choices as consumers in a disposable economy have made our everyday lives monuments to waste. Look in your bathroom: The shampoo, body wash, cosmetics, shaving cream, deodorant and health products you buy come in packages that cost us three times as much as the manufacturing cost of the products they hold. In other words, the hair conditioner bottle destined for the trash is actually more costly and valuable than the hair conditioner it contains. And these disposable things are designed to have a few days' useful service, then survive for thousands of years as potentially harmful and definitely costly trash. Why do we tolerate that?

In 2011, Americans spent $11 billion on a record 9.1 billion gallons of bottled water of equal or lesser quality than tap water that costs consumers ten thousand times less. Twenty-five to 50 percent of that expensive bottled water turns out to be mere tap water put in a plastic bottle with a fancy label slapped on, not to mention a bigger markup than any other product this side of the Hope Diamond. We throw out 60 million water bottles a day, in a country where the high quality and safety of inexpensive tap water is the envy of most countries in the world. One full day's worth of America's total oil consumption—about 18 million barrels—is spent hauling that bottled water around. Why?

The waste picture for milk is even worse than bottled water: Thirty percent of the milk produced in America is thrown away because of inefficiencies that let it expire or spoil before it can nourish anyone. All that energy, all that shipping, all that cattle feed, all that refrigeration, all that effort—nearly a third of it is wasted, thrown away, trashed. The rest of our food supply suffers similar losses—in a world where hunger is a growing, deadly problem, at least a quarter of the total American food supply is fated to become garbage. And we're paying for that, every day—in higher food prices, high utility bills, pollution and debt.

We accept products, from phones to stereos to televisions, designed to be disposed of rather than repaired or upgraded. We have transformed soda from a treat to a staple, though it has negative nutritional value and wastes colossal amounts of plastic, petroleum and water while fueling an unprecedented wave of child and adult obesity. The added bonus: Recent research has found that a common plastic chemical, BPA, that can leach from plastic beverage bottles and is found in most of our systems, may make the human body further prone to obesity. We drive cars so primitive in their design that they waste four-fifths of the energy produced by burning gas. Our power plants use only a third of the energy produced by burning coal, with the rest wasted, quite literally going up in smoke. Did you know that the average cable TV box plugged into your wall, a device that never fully shuts down even when your television is off, uses more electrical power than most refrigerators? Why does your cable company accept such a ridiculously wasteful device from the manufacturer? Why does your city council that contracts with the cable company accept such a system? Why do you accept it, since you're the one who has to pay the waste-bloated utility bill?

It seems such waste has a constituency of its own. Consider junk mail. Half the U.S. mail is now junk—or, to use the postal service's innocuous-sounding term, "direct mail." Americans received 85 billion pieces of the unwanted stuff in 2011. Those flyers and credit card offers and official-looking envelopes offering adjustable rate mortgages from The Loan Doctor collectively weigh in at 4 million tons—about one in every 100 pounds of trash headed to the landfill.

Junk mail is generally considered an unwanted annoyance, harder to shake than a cold. Yet taxpayers subsidize and encourage it without even realizing because junk mailers receive a postal rate far lower than a first-class stamp. Then the junk mailers receive a second subsidy because they are excused from cleaning up the tidal wave of waste they create. Consumers and taxpayers pick up the tab for that as well. Junk mailers, in short, have double incentives to be wasteful, two big helpings of corporate welfare that no other developed nation tolerates.

Such perverse incentives for waste permeate the economy. Most sanitation systems charge homeowners the same rate for large amounts of trash rolled to the curb as they do for small amounts—one flat fee for all, whether your neighbor makes half the trash you do, or twice as much. But some communities use a "pay as you throw" model: make less waste to be hauled away, use a smaller size bin at the curb, and you pay less each month. Bigger trash bins receive bigger bills because there's more to haul—an eminently fair setup. With that model, an incentive to be wasteful is replaced by an incentive to be thrifty. Give each homeowner a recycling bin and make hauling its contents free regardless of the amount of recyclables inside, and another incentive is born: an economic incentive to sort trash properly (which a surprising number of people resist under the *what's in it for me?* objection to the minor inconvenience of sorting).

According to a slew of EPA studies, pay-as-you-throw towns send an average of 40 percent less waste to landfills than other communities. It's a fairer system, it works, it provides the most beneficial incentives, and it has been proven to reduce garbage volume dramatically—yet fewer than one in five communities in the U.S. do

it. Changing wasteful ways, even when doing so is simple and essentially cost-free, is hard. Inertia and habit and fear of change get
in the way.

"MANAGING" WASTE is universally viewed as a positive. Everyone
wants clean cities, sidewalks and streets, a healthy, sanitary environment for our kids. But our focus on managing a waste problem
by making it appear to disappear has blinded many of us to the
reality of how much food, fuel, water and other things of value we
waste every day. For most of human history, such waste has been
viewed as shameful or worse. Gluttony is, after all, one of the seven
deadly sins, and it's not because it's associated with obesity, a threat
to an individual's survival, but because it represents overconsumption to the point of wastefulness, a threat to an entire community.
Today, however, a gluttony of consumption has become the norm.
That postwar marketing ploy of J. Gordon Lippincott, the push for
us to throw perfectly good things away and buy new things to replace them so that somebody else can get rich—an idea that goes
against our basic instincts and common sense—still holds us in
thrall. We are married to a disposable economy dependent on waste.

This really didn't make much sense in times of plenty. It certainly makes no sense today. The challenge for all of us is to find the
way back. It's a good time to stop managing waste, and start wasting less.

Which brings us to the coolest thing about trash, and the most
heartening thing about our horrifying 102-ton legacy: It is one of
the few big societal, economic and environmental problems over
which ordinary individuals can exert control. You don't have to fight
City Hall to do it. You don't have to organize protests or marches or
phone banks or political action committees. As a consumer, as a

homeowner or renter, as a person who eats and wears clothes and drinks water, you can choose to be more or less wasteful. You can choose to save more and spend less, which automatically means you will waste less. You can ban the bag from your own daily life. The smallest of steps can shave a piece from those 102 tons and save money for your household while you're also saving the planet. Bea Johnson sets an amazing example that can be daunting to the rest of us, but remember, it took her family two years to transition to a low-waste lifestyle. Not all their choices are right for other families, nor do they have to be in order to go on a useful trash diet. It can start small, a slow shift to a new normal. Little changes that, if they go viral, will carry big payoffs.

That's my challenge. I'm going to suggest five things anyone can do to be less wasteful. Try them out. Then suggest five of your own. E-mail them to garbology.book@gmail.com or post them on Twitter @EdwardHumes and we'll start a conversation about figuring out the best strategy for making America less trashy and Americans a bit richer in the process.

Here are my five:

1. **Refuse.** Bea Johnson's simple decision to just say no to a lot of stuff is the home run of waste reduction. From unwanted mail-order catalogs to recreational shopping excursions to printed phone bills rather than virtual ones, just refuse them. Say no to those stupid promotional key chains and tchotchkes that come free at conferences and fundraisers. You know it's junk, and accepting it just encourages more. Refuse. Your trash pile will shrink dramatically.

2. **Go Used and Refurbished.** Whether it's a computer, a TV, a car, a book or a coat, used or refurbished goods are always cheaper, are often indistinguishable from new (and many manufacturer-refurbished computers even carry same-as-new warranties), and

their environmental footprint is a fraction of that of new products. You are keeping resources out of the waste stream and saving yourself big bucks, all at the same time.

3. **Stop Buying Bottled Water**. It's a waste and a fraud wherever domestic water supplies are safe, which is virtually everywhere in the U.S. You don't need it. Get a couple of reusable bottles and put tap water in them.

4. **No More Plastic Grocery Bags**. No, this one won't save the world (though it will help the oceans), but Andy Keller's right: Plastic bags are the gateway drug of waste. If you can get that monkey off your back, you'll see how easy it can be to start chipping away other parts of your 102-ton legacy.

5. **Focus on Cost of Ownership**. The disposable economy wants you to think about the cost of things at the checkout stand. That's how we end up with cheapo DVD players that become trash in a year, clothes that fade and wear out after a few washes, and cable boxes that eat more electricity than a fridge. The disposable economy gives us things that are cheap in the short term but costly and wasteful over time. Saving up for purchases of things that are more durable, long-lived, reliable and efficient saves money over time, and radically reduces the waste we produce. And it does something else: The act of saving for something that's really good, something that we really want in our lives for years to come, encourages us to say no to other things we don't really need. It encourages saving instead of spending. And that means far less waste, too.

CUTTING WASTE, be it in government, in business or in the home, always makes sense—economically, environmentally and morally. It is a strategy that always has benefits. Waste is tied to all the big problems of the day, from climate change to peak oil to high energy costs and rising prices of the raw materials our industries and infrastructure require. Waste-cutting is the secret to sustainability, security and prosperity. That 102-ton legacy doesn't have to be the end of the story. It's in everyone's power to make it the starting point instead.

Send your top five (or two or ten) waste-cutting solutions to garbology .book@gmail.com, or post them on Facebook.com/Garbology, or send them to Twitter @EdwardHumes. Let's crowd-source the 102-ton legacy into oblivion . . . or at least put it on a major diet.

AFTERWORD

When the Los Angeles City Council convened in spring 2012 to consider the environmental cause of the moment, it was not climate, smog or solar energy on the agenda. The council had assembled to consider a ban of the lowly plastic shopping bag.

Bag partisans packed the cavernous, theater-sized chamber, and a noisy circus, also known as the council's public comment phase, ensued. Celebrities and plastic-bag assembly line workers offered dueling narratives of plastic pollution in the oceans and plastic job loss in the workplace. Paper bags were vilified as bigger polluters than plastic, plastic bags were vilified as the "gateway drug" for addiction to wasteful disposable goods, and each side accused the other of peddling junk science. An industry representative warned that all this fuss about zero waste would end up creating zero growth. Disabled veterans who had launched a reusable bag start-up testified otherwise: They said their product created more jobs than the disposable bag business ever had. Meanwhile, an angry, pro-bag homeless woman ("Yeah, I'm a bag lady, now shut the f—k up!") tussled with actress and anti-bag activist Sharon Lawrence, who looked a lot more flustered than she ever did back when she played Andy Sipowicz's prosecutor beau on the classic nineties TV show *NYPD Blue*. A real cop wearing LAPD blue finally had to intervene.

Through it all, a jarringly modern digital timer flashed at the front of the ornately wooded council chambers, limiting speakers to thirty-second snippets. This is the council's great democratizer, as it makes no distinction between the raving conspiracy theorist, the plastic industry lobbyist and the marine plastic pollution expert who raced to the meeting moments after his research vessel docked from Japan. Each got the same, brief chance to fast-talk their messages. And each was equally ignored by the fourteen council members present, who shuffled papers, issued orders to aides and whispered among themselves (except for the period of attentive star-powered solicitude accorded former *Seinfeld* actress Julia Louis-Dreyfus's extended turn at the podium).

Such institutional rudeness and favoritism is typical of circus time, but the comments mattered even less than usual: The hearing's outcome was all but preordained. With little fanfare, debate or angst, the council voted 13-1 to make Los Angeles the largest city in America to ban plastic shopping bags.

The circus time had been so long and unintentionally comic that many of those present seemed to miss the full import of the vote: that a group of politicians, whose time and energy is usually spent trying to offend the least possible number of voters, had acted decisively to tackle an issue that has been controversial, even incendiary, in many other communities. In the far less populous California county of San Luis Obispo to the north, angry citizens nearly launched a secession movement over a bag ban that same spring. Yet L.A.'s leaders matter-of-factly declared that plastic grocery bags should be a thing of the past in a city of nearly 4 million people. The most ubiquitous and recognizable disposable product in the world, used multiple times a day by the vast majority of the council's constituents, had been declared extinct within the nation's second largest city with barely a blink.

And it seemed that bags might be just the beginning. At the conclusion of the meeting, one councilman, Bill Rosendahl, held aloft a single-use water bottle—another immensely popular disposable product—and made a promise. "These are next," he shouted, to a round of applause.

It can be argued, quite correctly, that plastic bags are only a half of one percent of our trash, which means bag bans barely register when it comes to America's dirty love affair with trash—unless they are part of a larger strategy. Shifting a community, a country, an economy and a way of life so grounded in disposable products, in cocoons of packaging, in the "convenience" of wastefulness, in plastic-wrapped everything, seems a daunting prospect at best. Going on to the next disposable thing, whether water bottles or plastic foam or something else, will be more difficult still, and then there's the next thing, and the next after that. Such sweeping change, it is said, will come hard, if at all.

Or perhaps it's not as hard as we tend to think. A year before the city of Los Angeles passed its new bag law, the County of Los Angeles imposed its own ban on plastic bags and a 10-cent charge for paper sacks for stores outside the L.A. city limits. The move stirred great public consternation. People missed their plastics. They were upset by the inconvenience. Large numbers of shoppers simply ignored the ban or forgot about it. Then they ended up in line at the cash register during the first days of the ban without their own bags, yet unwilling to pay for paper sacks that previously had been free. Disgruntled shoppers huffed and staggered from the checkout stand, cradling bottles and baguettes and sacks of rice in their crossed arms. Some saw their cans slip from their grasp in the parking lot, thump to the asphalt and roll under cars as shoppers performed odd little jigs to avoid crushed toes.

All told, L.A. County's assault on the plastic bag—and the larger disposable, single-use economy it symbolizes—had a most inauspicious launch. And the opponents of such bans took heart. Public disdain for cost and inconvenience is a potent weapon in the battle of the bag. If people got angry enough, that would be the end of it.

But in July 2012, after the ban had been in place for a year, the California Grocers Association reported something remarkable. There had been a 75 percent reduction in disposable bag use. L.A. County officials put the figure even higher: 95 percent. Either calculation represents a somewhat stunning and unexpected transformation. Reusable bags had become the norm in Los Angeles County in just twelve short months.

How hard is it, then, to get an entire community to abandon long-held wasteful ways? The L.A. bag ban put a price tag on it: Apparently all it takes is one thin dime.

A YEAR after the publication of *Garbology*'s first edition, waste— and its elimination—has moved to the forefront of environmental concerns. Nearly one hundred communities around the country took action against disposable plastic and paper bags, either adopting bans, or putting them on the agenda for consideration and study. The entire state of Hawaii enacted a plastic bag ban in 2012 to combat the island state's horrendous trash and plastic pollution problems. San Francisco amended its breakthrough 2007 ban on bags, no longer limiting it to major supermarket and drugstore chains, but expanding the ban to include all retailers and restaurant takeout, while also imposing a dime charge for every paper grocery bag. The goal was to provide consumers with a strong incentive for choosing reusable over disposable bags of any type. Because this wasn't a tax—shop owners get to keep those dimes—the state grocery retailers association supported the city's move, and the ordinance survived the inevitable court challenge (though the industry is appealing).

Beyond bag bans, some communities began doing what the Los Angeles city councilman merely threatened: taking on the $11-billion-a-year bottled water industry, which has far more money, clout and customer loyalty than plastic bags. Even so, Concord, Massachusetts, famous for its pivotal role in the American Revolution, became the first city in the country to ban outright the sale of single-serving plastic water bottles of one liter or less. A first offense leads to a warning to the seller, a second nets a $25 fine, and the city will charge $50 per bottle sold after that, starting on January 1, 2013. Selling, not possessing, is the offense, and the ordinance exempts times of emergency. Championed by an eighty-four-year-old Concord resident, Jean Hill, who argued that bottled water was a waste of money, the city ordinance relies on an obscure 1981 U.S. Supreme Court decision that upheld a ban on non-returnable, non-refillable milk containers adopted in Minnesota to curtail waste.

Meanwhile, fifteen colleges and universities in the U.S. and Canada moved to adopt similar water bottle bans on campus, including Harvard's School of Public Health and Loyola University in Chicago. Four major municipalities—New York, San Francisco, Cook County, Illinois, and Seattle—banned the use of government funds to purchase single-use water bottles.

The bans and limits on bags and bottles aren't merely an attack on the disposable economy. These moves also reflect a dawning realization that recycling, while important and useful, is not a panacea. A preferable tactic might be to create less waste in the first place, saving money, resources and the environment on the front end of our consumer economy instead of the back end. "People," says ChicoBag's Andy Keller, "are starting to think differently. They feel moved to act, to do something positive."

Certainly this was reflected in the outpouring of suggestions and ideas sent in response to the invitation at the end of *Garbology* to "crowd-source" waste solutions. Reader suggestions ranged from the inspirational to the practical. Holly of Los Angeles, for example, urges everyone to start with simple changes:

- Invest in a Thermos and a water bottle
- Say no to plastic bags
- Use cloth napkins
- Repurpose items before trashing whenever possible—kids love to make things with those cardboard tubes, bubble wrap, etc.

Rebecca of Texas writes: "As soon as an unwanted catalog arrives, call immediately to get off the list. I find those sites that offer to get you off catalog lists ineffectual. I call or mail solicitors to ask to get off their lists also . . . but you gotta do it right away (easier when you are retired!)."

Jack, a fifteen-year-old high school musician in Sarasota, Florida, thinks educating people about waste is of paramount importance. "My band, The Garbage-Men, tries to raise awareness on recycling and reuse by playing hit songs on instruments we made ourselves out of recycled materials and reused garbage. My guitar is a Peanut Butter Captain Crunch cereal box with a broken yardstick neck and toothpicks for frets. It also has an electrified bottle cap for a pickup and a Disney shampoo bottle for a bridge! All this stuff was either in the garbage or headed there."

Liz of San Diego has seven waste-cutting tips that begin with an emphasis on charity and sharing:

- Donate scraps of ribbon, bottle caps, corks, pretty canceled stamps, lace, net, and any small bits of glittery material to a children's art studio. In San Diego we have the Rare Hare Studio on Adams Ave., and every time I find myself in that part of the city I drop off a bag.
- Bring my own "to go" box when dining out if it is not possible to share a meal with my companion.

- Bring my own mug and cutlery and cloth napkin to meetings where I know that coffee and food will be served.
- Personally call the sender of every unwanted bit of U.S. mail and ask to be taken off their list. Sometimes I give my e-mail address if I think I might want to hear from them again. I no longer receive any catalogs and some days no mail at all.
- Bring my own soap in a container when traveling and avoid using the small packages of toiletries found in every hotel room. After a single use they must wind up in some city's landfill.
- Use an old-fashioned ink pen instead of a ballpoint. Mine is greater than 50 years old, and I fill it with liquid ink using an old glass syringe. (This cannot be taken on airplanes because of leakage from pressure changes.)
- Nourish my acid-loving plants with coffee grounds.

"I may not be the first person to do the above," writes Liz, "but they are changes that I have made in my own behavior to decrease my 102-ton legacy."

Nancy of Eugene, Oregon, finds that people often harbor inhibitions about fixing things. She writes: "Get over it! Start learning how to take care of things! I can't tell you how often I pick up free items, oftentimes not cheap items, that people put out for free on their curbs *just* because there was a loose screw (literally!)." She also suggests returning the favor by giving away, rather than throwing away, unwanted items to charities or listing them as free on such websites as craigslist—even items in need of repair. "You'd be amazed at what people will reuse and even buy if given the chance." A trend of increasing numbers of pianos turning up at landfills around the country because of the inconvenience of donating such treasures horrifies her and others, including Barbara of Dallas, who likened such wasting to sin. "Think of all the churches or youth groups or schools that would benefit!"

Val of New York City has a number of suggestions, but at the top of her list is confronting her love of buying stuff, which turned out to be a common thread in the *Garbology* mailbag—the mall as a kind of addiction, even in a difficult economy. "For me," Val writes, "that means cutting down on that recreational shopping." Leah of Carmel, California, has a possible solution for Val: "Re-gifting," which she finds "both enjoyable and wasteless."

Jorge of Miami thinks the best path to lower waste could be through tax breaks for manufacturers who reduce packaging and single-use products, and who establish meaningful take-back programs for recycling and refurbishing rather than landfilling. "We have to prime the pump to close the loop," he writes.

There were too many suggestions to publish all of them, but the note from Maria of Garwood, New Jersey, is a must-read—a reminder that the next generation just might clean up our mess for us, particularly with teachers like Maria to show them the way:

> I teach preschoolers and for the last four years I have encouraged the children (really, the parents) to use reusable containers for snack. I started a worm food container and have a compost bin for their banana peels, apple cores, etc. I praise a child when they have zero garbage (for Earth Day, we compare our garbage with another classroom to show how little we have!). When they leave to go to kindergarten, I give them handmade reusable snack bags and I hope I am sending them off thinking about the environment. I have had moms and dads tell me their children reprimand them if they don't recycle or properly dispose of something. And

they tell me their children are still conscious of the environment even though they have been out of my class for several years. I hope this teaching someone else to do the right thing gets my 102-ton legacy down even further!

I want to thank all the *Garbology* readers who shared their trash-busting, zero-waste tips and practices. You have shown that waste really is the one big problem anyone can do something about, and how each one of us can do better. I invite you to keep this conversation going by connecting with the *Garbology* Facebook page. It's a daily discussion of how we can help our economy, our environment and future generations by refusing, reusing, repurposing and recycling—how, as Maria put it, we can all learn to do the right thing to shed that 102-ton legacy.

ENDNOTES

INTRODUCTION

1. This calculation is derived from the most recent and most accurate data on America's annual municipal waste generation, the biannual study by Columbia University and the journal *BioCycle*, which put the nation's trash total at 389.5 million tons in 2008. The population of the country was put at 301 million that year by the U.S. Census, which yields a daily waste generation amount of 7.1 pounds per day.
2. "Plastic Water Bottle-Makers Sued by California over Green Claims," *Los Angeles Times*, October 27, 2011.
3. "Products, Packaging and US Greenhouse Gas Emissions," Joshuah Stolaroff, Product Policy Institute, September 2009.

4. "The State of Garbage in America," a joint study by *BioCycle* and the Earth Engineering Center of Columbia University, by Rob van Haaren, Nickolas Themelis and Nora Goldstein, published in *BioCycle*, October 2010. Data is from the year 2008. The study is published biannually.

5. This calculation assumes a U.S. adult population of 230 million and an average weight of 178 pounds (195 pounds for men and 165 pounds for women), as reported by the National Center for Health Statistics in "U.S. Body Measurements, 2009."

6. The *BioCycle*/Columbia University biannual survey of municipal solid waste sent to landfills, recycling, compost and waste-to-energy facilities draws on actual state-by-state data from the nation's municipal waste systems and is the most accurate actual count of America's trash. The better-known annual MSW report from the EPA does not use actual trash disposal data, but instead relies on a materials flow analysis and data from manufacturers to estimate the amount of products and materials consumed by Americans and how long those products and materials are likely to last. From these assumptions, combined with waste characteristic sampling studies for non-manufactured waste, the EPA estimates calculate how much stuff ought to be thrown out every year. Actual trash data is not used by the EPA. This method has come under fire for its chronic tendency to underestimate total trash and landfill loads, while overestimating the proportion that gets recycled.

7. Ibid.

8. Garbage In, Garbage Out: A Note on the Numbers

The U.S. Environmental Protection Agency's annual report, "Municipal Solid Waste in the United States," is widely considered the most authoritative source on waste and trash in the country, a garbage ground zero for journalists, researchers and elected officials on how much trash we make, burn, bury and recycle, and how much of it is plastic, paper, metal, food scraps, or yard trimmings. Overall, according to the EPA, the country's annual "waste stream" broke down in 2008 this way: 54 percent of the municipal waste (135.6 million tons) went to landfills, a third (84 million tons) was recycled or composted, and the remaining 12.6 percent (31.6 million tons) was burned in waste-to-energy generating plants. The grand total of municipal waste reported: 251 million tons. At that number, America's daily trash footprint would be 4.5 pounds a person. But that's more than 2.5 pounds a day less per American than the correct amount, 7.1 pounds, and more than 130 million tons light for the whole country's yearly tally.

So how can that be? Where did the EPA go so badly wrong with a report it's been producing for decades?

Most might guess coming up with trash numbers would involve a lot of weighing of the streams of trash headed to landfills. This would be a relatively straightforward task—laborious, but straightforward. Every municipal waste landfill in America has scales. They weigh garbage trucks going in full, they weigh them going out empty, and by calculating the difference, they determine how much trash gets dumped—each load, every load, every day of the week. It's how dump operators plan for the future, budget their resources and manpower, and, not incidentally, it's how they make money: They charge by the ton. Recycling, composting and waste-to-energy operations work in an analogous way to produce a statistical snapshot of our waste. Many states compile reports summarizing this data in order to plan and evaluate their own conservation and recycling efforts.

But the EPA does not use this information. It does not weigh trash in the real world—not a single piece of it—nor does it contact the nation's landfills to get that information. Instead, the EPA relies on "materials flow methodology." In plain English, this means the EPA calculates trash amounts based not on objective weights and measures, but on data supplied by manufacturers on how much stuff they sell (for instance, the number of plastic bags made and sold in the U.S. every year), how long that stuff is likely to last before becoming trash, and how much of it gets recycled, composted or burned. These are industry estimates reported through a national honor system, checked by equations, not scales. Waste sampling studies are then used to estimate national figures for yard trimmings, food scraps and other non-manufactured municipal waste. Sometimes press reports on garbage are used to flesh out the data further. Together, this amalgam of information is used to produce an estimate of the total waste stream—a figure lying at the end of a long chain of promises, assumptions and theory.

This method dates back thirty years, to an era when there were ten times the number of landfills and thousands of illegal dumps in the U.S., and the industry was largely unregulated and uncharted. Using the indirect method of materials flow analysis made sense then— it was the best anyone could do. But there are far fewer landfills now, a web of state reporting requirements have been placed on them, and the ability to do a direct, more accurate count of waste, rather than rely on indirect life-cycle calculations, has existed for more than a decade.

The flaws in the EPA's approach are easily detected. The EPA estimates that a total of 135 million tons of trash were buried in landfills in 2008. The problem: A single landfill operator, Waste Management, Inc., reports burying almost the same amount of trash that year, 125 million tons, all on its own. Waste Management may be the biggest trash company in the world, but they don't own America's entire landfill business—they control only a third of America's active landfill space. There are more than a hundred other major waste-management companies in the country, not to mention the many publicly owned and operated landfills, and their combined landfill business easily exceeds Waste Management's. One simple check reveals that the EPA numbers are badly off-kilter.

It fell to a partnership between Columbia University's Earth Engineering Center and a respected, if obscure, trade journal, *BioCycle*, to do the actual trash counting that the feds had declined to do. This project produces numbers from the real world of trash that reveal the serious, even scandalous, gap between the EPA stats and reality—the biggest, dirtiest and poorest-kept secret in the trash biz.

How bad is the disparity? Americans are sending more than twice as much garbage to municipal landfills as the EPA figures suggest. Adding insult to injury, the EPA also incorrectly inflates the proportion of trash recycled—we're not doing nearly as well as we thought. The amount recycled and composted isn't a third of all our trash, as the EPA reported for the last several years. It's barely a quarter of it. In 2011, the EPA leadership finally admitted there was a problem and publicly solicited advice for improving its annual garbage survey.

Not all EPA solid waste statistics are flawed, however. While the materials flow methods used to calculate the amount of trash aren't working well, the methods used to calculate the *composition* of our trash continue to be useful. These calculations are informed in part by studies of real-world samples of typical Americans' trash—how much of it is plastic, metal, paper, food scraps and so on. These figures are expressed in the EPA annual reports as percentages. Because extrapolating national estimates from real-world samples is a tried-and-true, scientifically valid technique, the EPA's percentage estimates on the composition of trash are used throughout this book as the best available data. However, in passages or lists in which those percentages are used to derive quantities of a certain type of trash, such as reporting that 5.4 million tons of rugs and carpets are sent to landfills each year, this quantity is calculated by applying the EPA's composition percentages to the Columbia/*BioCycle* total waste figures.

9. Even the Pentagon sources its silicon from the same China that, as recently as 1999, was banned from importing Apple Inc.'s most powerful personal computer because it might be used in weapons systems. (Of course, ten years later, most Apple products, like every other U.S.-branded computer, tablet and smartphone, were being built in Chinese factories.)

10. *Journal of Commerce*.

CHAPTER 1

1. "The State of Garbage in America," *BioCycle*, October 2010.

2. "Mission 5000," EPA, 1972.

3. The Comprehensive Environmental Response, Compensation and Liability Act of 1980, better known as the Superfund, is a federal program for cleaning up hazardous waste sites. It was created in response to the

severe pollution and health threats posed by the Love Canal disaster and other, similar crises. At the end of 2010, there were 1,280 sites slated for cleanup on the Superfund priority list.

4. According to Seagull Control Systems, Inc., which markets such seagull barriers. The U.S. Department of Agriculture also recommends monofilament lines as the most effective safeguard against landfill-marauding gulls.

CHAPTER 2

1. From *Municipal Journal*, Volume XLV, No. 26, July–December 1918; and "Health Survey of New Haven: A Report Presented to the Civic Federation of New Haven by Charles-Edward Amory Winslow, James Gowan Greenway and David Greenberg of Yale University," Yale University Press, 1917.

CHAPTER 3

1. "Consumption of Sugar Drinks in the United States, 2005–2008," National Center for Health Statistics, August 2011.

CHAPTER 4

1. "Mission 5000," EPA, 1972.

CHAPTER 10

1. Estimates of plastic bag usage by the average American vary. Industry estimates put the figure at five hundred disposable bags per capita, according to American Plastics Manufacturing, Inc. Andy Keller of ChicoBag considers this to be a very conservative figure, and has published calculations based on manufacturing and EPA data that peg annual plastic bag disposal for 2009 at 739 bags per person. This figure may also be too low, as the EPA consistently underestimates the amount of trash generated by Americans.

2. Keller's calculation is based on the annual consumption of plastic grocery bags in the U.S. reported by the International Trade Commission

in 2009—102 billion—multiplied by his assumed average bag length of 1 foot, then divided by the earth's circumference, which is 131.48 million feet.

3. International Trade Commission, 2009.

4. American Plastics Manufacturing, Inc.

5. "By 'Bagging It,' Ireland Rids Itself of a Plastic Nuisance," *New York Times*, January 31, 2008.

6. Substituting a local bag ban for a bag tax was the brainchild of San Francisco Supervisor Ross Mirkarimi, a quixotic Bay Area politician whose spectrum of achievements defies categorization. He graduated president of his class at the San Francisco Police Academy, worked for the district attorney investigating white-collar crime, cofounded the California Green Party and supported the legalization of marijuana. Mirkarimi argued that taking on the plastic bag "plague" was a necessary first step in healing environmental damage: "Instead of waiting for the federal government to do something about this country's oil dependence, environmental degradation or contribution to global warming, local governments can step up and do their part. The plastic bag ban is one small part of that."

7. California Secretary of State, lobbying activity reports.

8. "Miracle No: Groups Call on Major Leagues to Denounce Miracle-Gro Deal," SafeLawns.org, March 14, 2010.

9. "Recall of Scotts Miracle-Gro Products," EPA, Region 5 Pesticides, May 2008.

10. "Small Plastic Bag Lawsuit Could Have a Huge Impact on Green Business," *Forbes*, June 21, 2011.

11. "Battle of the Bags," BagMonster.com, September 25, 2011.

12. "Plague of Plastic Chokes the Seas," *Los Angeles Times*, August 2, 2006.

13. "What We Actually Know About Common Marine Debris Factoids," FAQ, NOAA Marine Debris Program, http://marinedebris.noaa.gov/info/faqs.html#5.

14. "Hilex Poly and ChicoBag Reach Settlement over False Marketing Claims," PR Newswire, September 13, 2011.

15. "Advance Polybag and Superbag Support Hilex Poly's Victory in Settle-

ment of ChicoBag Suit," press release transmitted by Reuters, September 15, 2011.

16. "An Overview of Carryout Bags in Los Angeles County," staff report to the Los Angeles County Board of Supervisors, August 2007.

CHAPTER 11

1. "How Denmark Paved Way to Energy Independence," Leila Abboud, *Wall Street Journal*, April 16, 2007.

2. Amagerforbrænding website, http://amfor.dk/English/Incineration.aspx.

3. "Is It Better to Burn or Bury Waste for Clean Electricity Generation?" *Environmental Science & Technology*, vol. 43, November 6, 2009.

INDEX

ABOUT THE AUTHOR

EDWARD HUMES is the author of eleven critically acclaimed nonfiction books, including *Force of Nature, Monkey Girl, Over Here, School of Dreams, No Matter How Loud I Shout* and the bestseller *Mississippi Mud*. He has received the Pulitzer Prize, the PEN Award and numerous other awards for his journalism and books. He has written for the *New York Times*, the *Los Angeles Times*, *Los Angeles Magazine* and *Sierra*. He lives with his family—including the two most recent additions, a pair of rescued racing greyhounds—in California.